Segment Routing in MPLS Networks

Transition from traditional MPLS to SR-MPLS with TI-LFA FRR

Hemant Sharma

‹packt›

Segment Routing in MPLS Networks

Group Product Manager: Dhruv J. Kataria
Publishing Product Manager: Prachi Sawant
Book Project Manager: Uma Devi Lakshmikanth
Senior Editor: Roshan Ravi Kumar
Technical Editor: Arjun Varma
Proofreader: Roshan Ravi Kumar
Copy Editor: Safis Editing
Indexer: Rekha Nair
Production Designer: Alishon Mendonca
DevRel Marketing Coordinator: Rohan Dobhal

First published: November 2024

Production reference: 1241024

Published by Packt Publishing Ltd.
Grosvenor House
11 St Paul's Square
Birmingham
B3 1RB, UK.

ISBN 978-1-83620-321-6

www.packtpub.com

I am sincerely grateful to the individuals and resources that have significantly contributed to my learning and, ultimately, to the creation of this book. Without their invaluable input, this project would not have come to fruition.

I consulted a diverse range of materials throughout the writing process, both online and offline. While not explicitly mentioned here, their collective contributions were indispensable in developing the substance and depth of this work.

To all these sources of knowledge, I offer my deepest thanks for their generosity in sharing their expertise, which has been instrumental in shaping and enriching this book.

– Hemant Sharma

Foreword

If you choose an MPLS-based WAN architecture, **Segment Routing MPLS (SR-MPLS)** solves many problems and provides several benefits: less control plane overhead, better use of the available link bandwidth, and fast rerouting around failures at a global scale. Sounds almost too good to be true! However, this book can help enterprise and **Service Provider (SP)** network engineers build trust in SR-MPLS.

That trust begins with a much-experienced author, Hemant Sharma, who fits the model as an expert in the field. I first got to know Hemant via Cisco Champion's message boards, where his posts revealed his deep skills. Once we got to know each other, I learned more of his career story—an impressive journey. If you ever thought you'd love to have another experienced network engineer beside you to move into a new technology, know that you get that in Hemant and this book. In many ways, this book models what Hemant has seen work with SR-MPLS during his long travels in the SP engineering world.

Hemant's lab approach to the content made me most excited about the book. He wants you to build confidence and trust with SR-MPLS, and there's no better way than seeing it work for yourself. The book cycles through new learning followed by labs, with supplied configuration files you can use with EVE-NG and similar tools. Hemant weaves that feature throughout the book.

As for what you should expect to learn, he starts with some MPLS without Segment Routing as a baseline for all. Then, he lays out the core SR-MPLS concepts and features. The rest of the book then works through the variations on the biggest motivation to consider SR-MPLS: the various rerouting options available. If you ever thought about whether SR-MPLS makes sense for your network or how you might implement any of the various rerouting features, this book has you covered!

I expect we'll all see more from Hemant over the coming years beyond this first published book. He's active in the **Internet Engineering Task Force (IETF)**—it was his recent work with **BGP Monitoring Protocol (BMP)** that first caught my eye. His combination of real-world deep experience and his willingness to create organized and useful content is a wonderful development! I'll be watching this space!

Wendell Odom

CCIE Enterprise 1624

Author of 10 editions of the CCNA Official Cert Guide from Cisco Press

Contributors

About the author

Hemant Sharma is a seasoned network engineer with over 14 years of experience in SP networks. He holds a Bachelor of Engineering degree in information and technology and is a certified expert with credentials including CCNA, CCNP, and CCIE #28809 (Emeritus). An active contributor to the IETF, Hemant helps shape the future of networking standards. At Vodafone Group, Hemant leads the evolution of their Worldwide MPLS network and was instrumental in establishing the Vodafone Global SR-MPLS Network. He is recognized within Vodafone as an authority on IP/MPLS, MP-BGP, Multicast, and QoS. Hemant remains dedicated to staying updated with the latest networking trends, which underscores his reliability and expertise in the field.

Your exploration of the book is genuinely appreciated, and your feedback holds immense value. Should you identify any errors or areas for improvement, I extend my sincere apologies for any inconvenience caused. Your insights are crucial to the refinement process, and I am fully committed to addressing and enhancing the work accordingly. Your understanding and thoughtful engagement are greatly appreciated.

About the reviewers

Fernando Lopez Pajares is a senior network and security architect with over 16 years of experience in designing sophisticated network and security architectures. He holds a degree in telecommunications engineering and multiple certifications, including the prestigious CCIE (#54222). He has architected large enterprise networks and, at Fortinet, he provided strategic recommendations that guided the optimization of network and security frameworks for his customers. Currently, he is focused on security and is leading the security strategy and architecture at a large enterprise maintaining an interest in network systems, a field about which he is passionate.

Hari Vishnu, a distinguished senior consultant with a CCIE EI certification, is highly regarded for his expertise in designing and implementing data center fabrics utilizing application-centric infrastructure and VXLAN. His skills also extend to designing and commissioning ISP Core, Gateway, and Transport networks.

Shraddha Hegde has 23+ years of experience working for routing protocols in leading network equipment vendors. She is currently working with Juniper Networks. Her areas of expertise include ISIS, OSPF, segment routing, and MPLS technologies. She contributes actively to protocol standardization in IETF and has authored several drafts and RFCs.

Ron Bonica is a distinguished engineer at Juniper Networks, specializing in power management, IPv6, and segment routing. He is active in the IETF, having authored or co-authored 22 RFC documents and served three 2-year terms as co-director of the IETF Operations and Management Area. Ron currently co-chairs the IETF V6OPS Working Group.

Table of Contents

Part 2 - Segment Routing (SR-MPLS)

3

4

Part 3 - Fast Reroute in SR-MPLS Networks

5

6

Lab 5 – Zero-Segment FRR   137

7

Lab 6 – Single-Segment FRR   149

8

Lab 7 – Double-Segment FRR   161

11

Lab 10 – TI-LFA Local SRLG-Disjoint Protection   203

12

Lab 11 – TI-LFA Global Weighted SRLG Protection   215

13

Lab 12 – TI-LFA Node + SRLG Protection   229

14

Preface

The journey began with a spark of curiosity that ignited a deep desire to explore the subject firsthand. Eager to grasp the intricacies, I delved into hands-on practice with numerous labs, immersing myself in the practical aspects. As I honed my skills and knowledge, a unique idea emerged – to create a lab guide that could serve as a valuable resource for others venturing into this field.

The vision evolved further, as I realized the potential of consolidating all these efforts into a comprehensive book. Combining theory with practical experience, the book could offer readers a seamless learning path, catering to both their inquisitive minds and their desire for hands-on application.

The journey from curiosity to lab exploration and, finally, to crafting this book has been an exhilarating one. With passion and dedication, I aim to share this wealth of knowledge, hoping to inspire and empower others on their voyage of discovery and growth in this subject.

The primary objective driving the creation of this book is to facilitate a seamless transition from the traditional **Multiprotocol Label Switching (MPLS)** to the **Segment Routing MPLS (SR-MPLS)**. Recognizing the complexities and challenges that can arise during such a shift, the book endeavors to present the subject matter in a manner that simplifies the process for you.

By offering clear and concise explanations, practical examples, and step-by-step guidance, this book aims to empower network professionals and enthusiasts to adopt SR-MPLS with confidence. It seeks to demystify the intricacies and technicalities associated with the new approach, making it accessible and comprehensible to a wider audience.

The emphasis on simplicity in this book not only serves to ease the transition for experienced network engineers but also extends a welcoming hand to those new to this concept. Through well-structured content and user-friendly language, you are encouraged to embark on this journey with enthusiasm, knowing that you will be equipped with the knowledge and insights necessary to embrace the advancements in networking technology.

Overall, the book's elaborative approach ensures that you will gain a solid understanding of SR-MPLS, empowering you to implement it effectively and harness its benefits in your networking environments.

Who this book is for

This book is tailored for network professionals operating within MPLS environments, including network engineers, planners, designers, and architects. It particularly caters to those actively involved in SR-MPLS networks or currently in the process of integrating this technology into their network infrastructure.

Structured as a series of labs, the content of this book encompasses both theoretical concepts and practical knowledge. It serves as a comprehensive resource to enhance the understanding and management of SR networks, especially during the transition from traditional MPLS networks.

Upon completing this book, you will have the proficiency to understand, implement, and operate the fundamental elements of the following features within the Cisco IOS-XR network operating system:

- IS-IS IGP routing protocol
- MPLS LDP
- SR-MPLS in IS-IS networks
- TI-LFA in SR-MPLS networks

What this book covers

Chapter 1, Introduction to Multiprotocol Label Switching (MPLS), revisits the fundamentals of MPLS, providing you with a concise yet insightful review of its operations. Additionally, it offers valuable perspectives into the basics of segment routing.

Chapter 2, Lab 1 – Getting Started with LDP-Based MPLS Network, walks you through setting up the basic network topology. It covers using IS-IS as the interior gateway routing protocol and LDP as the MPLS forwarding protocol.

Chapter 3, Lab 2 – Introducing Segment Routing MPLS (SR-MPLS), introduces segment routing within an existing MPLS network based on LDP and explains how forwarding operates in both scenarios.

Chapter 4, Lab 3 – SR-LDP Interworking, concentrates on connecting different domains from LDP to SR-MPLS, and vice versa. You will understand how the inter-domain label switch path is created and the essential requirements for it. You will also grasp the significance of SRMS in facilitating interaction between SR-MPLS and traditional MPLS networks.

Chapter 5, Lab 4 – Introducing TI-LFA (Topology Independent – Loop-Free Alternate), introduces TI-LFA in SR-MPLS networks, detailing the calculation and installation of backup paths on routers.

Chapter 6, Lab 5 – Zero-Segment FRR, explores scenarios where backup paths can reroute traffic in the event of network failures, without any additional segments introduced.

Chapter 7, Lab 6 – Single-Segment FRR, examines scenarios where backup paths reroute traffic in case of network failures, requiring only one additional segment for rerouting.

Chapter 8, Lab 7 – Double-Segment FRR, examines scenarios where backup paths reroute traffic in case of network failures, requiring two additional segments.

Chapter 9, Lab 8 – Microloop Avoidance, discusses the occurrence of microloops during network convergence and explores their mitigation within SR-MPLS networks.

Chapter 10, Lab 9 – TI-LFA Node Protection, prepares the network topology for forthcoming TI-LFA scenarios. It specifically delves into the node protection method, where the backup path is computed under the assumption that a link failure implies a node failure.

Chapter 11, Lab 10 – TI-LFA Local SRLG-Disjoint Protection, explores the TI-LFA local-SRLG disjoint scenario, where the backup path is calculated under the assumption that all local SRLG links fail simultaneously.

Chapter 12, Lab 11 – TI-LFA Global Weighted SRLG Protection, builds upon the previous lab, examining scenarios where both local and remote SRLG links are bypassed in the calculation of TI-LFA backup paths.

Chapter 13, Lab 12 – TI-LFA Node + SRLG Protection, delves into a scenario where TI-LFA is tasked with computing and implementing a backup path that circumvents both local as well as remote SRLGs and the involved node.

Chapter 14, Lab 13 – TI-LFA Tiebreaker, addresses scenarios where TI-LFA is tasked with calculating both node protection and SRLG protection simultaneously. It explores how prioritization is determined when both protections can't be achieved concurrently.

To get the most out of this book

This book tackles a subject that is already extensively covered in articles, books, and videos. However, its unique approach aims to stand out by consolidating all the theoretical knowledge in one comprehensive resource. The central focus is on providing you with a seamless learning experience, supported by practical labs that enable smooth transitions between topics.

By presenting a well-organized compilation of theories and incorporating hands-on lab exercises, this book ensures that you can effortlessly navigate from one concept to another. It caters to both beginners and those with prior knowledge, offering a holistic understanding of the subject matter while facilitating quick and efficient learning.

Whether you are a novice eager to delve into the topic or a seasoned individual seeking to reinforce your expertise, this book serves as a valuable reference, condensing valuable insights from various sources into a single cohesive work.

Within the pages of this book lies a compilation of fundamental building blocks that form the bedrock of segment routing, presented at a rudimentary level. The intentional focus on foundational aspects means that not all the advanced features are covered in this edition.

The decision to limit the scope allows for a clear and coherent presentation of the essential concepts, making it accessible to readers at various stages of familiarity with the subject. However, the author acknowledges that there is still much more to explore and share about segment routing.

To fully leverage the labs featured in this book, it is imperative to have a solid command of the Cisco IOS-XR CLI. This foundational knowledge is complemented by familiarity with IS-IS and LDP running on the Cisco IOS-XR platform, as these technologies serve as the cornerstone for the practical exercises presented here. If you already possess prior experience with these technologies, you will find it more straightforward to follow along and apply the concepts in the labs.

For those who are new to the Cisco IOS-XR CLI, IS-IS, or LDP, there's no need to be concerned. The book is purposefully structured to provide detailed explanations and support, ensuring you gain the necessary understanding as you progress through the labs. So, whether you are well-versed in these technologies or approaching them for the first time, the book is designed to facilitate a comprehensive learning experience.

This book serves as a valuable resource for gaining foundational knowledge on the subject, but it is not intended to transform you into an expert. Instead, it provides a concise and approachable introduction, equipping you with essential insights and understanding. For those seeking comprehensive expertise, further exploration and additional resources will be necessary. Nevertheless, this book acts as a stepping stone towards a broader comprehension of the subject matter.

Download the example code files

You can download the example code files for this book from GitHub at `https://github.com/PacktPublishing/Segment-Routing-in-MPLS-Networks`. If there's an update to the code, it will be updated in the GitHub repository.

We also have other code bundles from our rich catalog of books and videos available at `https://github.com/PacktPublishing/`. Check them out!

Conventions used

There are a number of text conventions used throughout this book.

`Code in text`: Indicates code words in text, database table names, folder names, filenames, file extensions, pathnames, dummy URLs, user input, and Twitter/X handles. Here is an example: "The following output from the `PE1` router confirms that the IP addresses are applied according to the configurations on each router."

A block of code is set as follows:

```
router IS-IS IGP
interface GigabitEthernet0/0/0/3
address-family ipv4 unicast
metric 100
```

When we wish to draw your attention to a particular part of a code block, the relevant lines or items are set in bold:

```
interface {{ interface_name }}
description "P(E){{ x }}_to_P(E){{ y}}"
ipv4 address {{ xy.0.0.(x or y) }} 255.255.255.0
no shutdown
```

Any command-line input or output is written as follows:

```
RP/0/RP0/CPU0:PE1#show isis route detail 5.5.5.5/32
L2 5.5.5.5/32 [40/115] Label: None, medium priority
Installed Jan 01 06:02:54.493 for 00:02:56
via 12.0.0.2, GigabitEthernet0/0/0/2, P2, Weight: 0
src PE5.00-00, 5.5.5.5
RP/0/RP0/CPU0:PE1#
```

Bold: Indicates a new term, an important word, or words that you see onscreen. For instance, words in menus or dialog boxes appear in **bold**. Here is an example: "Built-in support for **SR-Traffic Engineering (SR-TE)**, allowing operators to direct traffic through specific paths for optimized network performance."

> **Tips or important notes**
> Appear like this.

Get in touch

Feedback from our readers is always welcome.

General feedback: If you have questions about any aspect of this book, email us at customercare@packtpub.com and mention the book title in the subject of your message.

Errata: Although we have taken every care to ensure the accuracy of our content, mistakes do happen. If you have found a mistake in this book, we would be grateful if you would report this to us. Please visit www.packtpub.com/support/errata and fill in the form.

Piracy: If you come across any illegal copies of our works in any form on the internet, we would be grateful if you would provide us with the location address or website name. Please contact us at copyright@packt.com with a link to the material.

If you are interested in becoming an author: If there is a topic that you have expertise in and you are interested in either writing or contributing to a book, please visit authors.packtpub.com.

Share Your Thoughts

Once you've read *Segment Routing in MPLS Networks*, we'd love to hear your thoughts! Please `click here to go straight to the Amazon review` page for this book and share your feedback.

Your review is important to us and the tech community and will help us make sure we're delivering excellent quality content.

Share Your Thoughts

Once you've read *Segment Routing in MPLS Networks*, we'd love to hear your thoughts! Scan the QR code below to go straight to the Amazon review page for this book and share your feedback.

`https://packt.link/r/1-836-20321-7`

Your review is important to us and the tech community and will help us make sure we're delivering excellent quality content.

Download a free PDF copy of this book

Thanks for purchasing this book!

Do you like to read on the go but are unable to carry your print books everywhere?

Is your eBook purchase not compatible with the device of your choice?

Don't worry, now with every Packt book you get a DRM-free PDF version of that book at no cost.

Read anywhere, any place, on any device. Search, copy, and paste code from your favorite technical books directly into your application.

The perks don't stop there, you can get exclusive access to discounts, newsletters, and great free content in your inbox daily

Follow these simple steps to get the benefits:

1. Scan the QR code or visit the link below

https://packt.link/free-ebook/978-1-83620-321-6

2. Submit your proof of purchase
3. That's it! We'll send your free PDF and other benefits to your email directly

Part 1 -
MPLS Overview
and Recap

In this part, you are provided with a comprehensive understanding of the workings of traditional MPLS networks. It incorporates background information on SR-MPLS and fast rerouting in MPLS. Additionally, the part delves into the network infrastructure, outlining the topology specifically designed for the book's context and elaborating on its fundamental building blocks.

This part contains the following chapters:

- *Chapter 1, Introduction to Multiprotocol Label Switching (MPLS)*
- *Chapter 2, Lab 1 – Getting Started with LDP-Based MPLS Network*

1

Introduction to Multiprotocol Label Switching (MPLS)

This chapter aims to explain **Multiprotocol Label Switching** (**MPLS**), an important technology in **Transmission Control Protocol** (**TCP**) / **Internet Protocol** (**IP**) networks, and its continued relevance. It also introduces **Fast Reroute** (**FRR**) in MPLS networks and **Segment Routing MPLS** (**SR-MPLS**), a newer method for operating MPLS networks. Before beginning the practical exercises, it explains the structure, the key components needed to build the required infrastructure, and where to obtain them. Throughout, you are encouraged to revise the fundamental concepts of MPLS, SR-MPLS, and FRR. Understanding these concepts is crucial for successfully following the practical exercises and gaining deeper insight into advanced networking principles in later chapters. By the end of this chapter, you will have understood MPLS concepts, their applications, and the required infrastructure for hands-on experience.

This chapter will cover the following topics:

- What is MPLS?
- Applications of MPLS
- Introducing FRR
- Introduction of **segment routing** (**SR**)
- Structure of the book
- Network infrastructure

Let us dive in!

What is MPLS?

In the TCP/IP model, the Internet Layer (Layer 3) is responsible for managing IP addressing and routing packets between different networks. MPLS sits between the traditional Layer 2 (Network Access Layer) and Layer 3 (Internet Layer) in the TCP/IP model, which is sometimes referred to as Layer 2.5 or Shim Header.

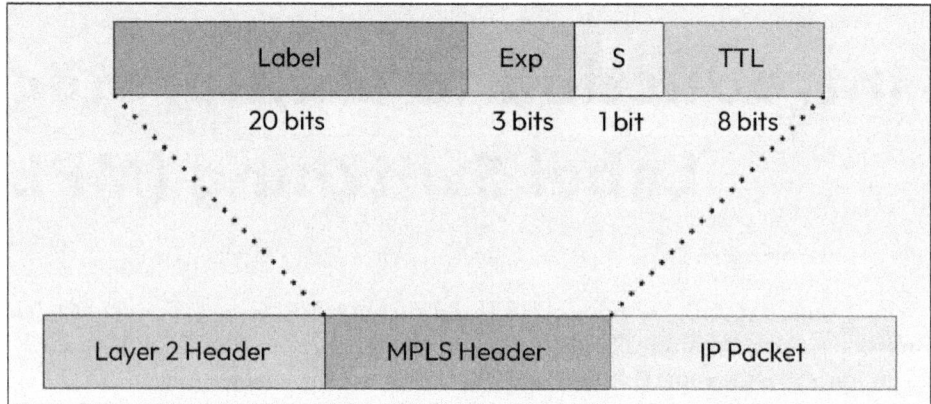

Figure 1.1 – MPLS Header

The MPLS header comprises four fields, totaling 32 bits in size:

- **Label**: The Label field is the most critical part of the MPLS header, comprising a fixed size of 20 bits that carries the forwarding information for the packet. Each router in an MPLS-enabled network maintains a forwarding table that maps incoming labels to the next-hop router and the corresponding output interface. The label is used to quickly determine the packet's path through the network, making the forwarding process more efficient.

- **Experimental (EXP) or Traffic Class (TC) bits**: The Experimental bits field is 3 bits long and is commonly used for implementing QoS in MPLS networks. By setting different values in the EXP/TC field, certain types of traffic can be prioritized over others, helping to ensure that critical or high-priority data receives preferential treatment in the network.

- **S (bottom of stack) bit**: This is a single bit that indicates whether the current label is the last one in the MPLS stack. MPLS supports label stacking, which means multiple MPLS labels can be used hierarchically. When the S bit is set to 1, it means this is the last label in the stack.

- **Time to live (TTL)**: The TTL field in the MPLS header functions similarly to the TTL in IP headers. It is an 8-bit field that is decremented at each MPLS hop. If the TTL reaches zero, the packet is discarded. This helps prevent packets from being trapped in routing loops.

The labs in this hands-on guide will focus on exploring MPLS labels, the label stack, and TTL (through the MPLS path) across various outputs. However, the EXP bit falls outside the scope of this book and will not be covered.

MPLS is a technology that transforms traditional TCP/IP networks through the introduction of a label-switching mechanism. This approach significantly improves the efficiency and scalability of network routing. Here's an overview of how MPLS changes TCP/IP networks with its label-switching mechanism:

- **Label allocation**: A label is a short, fixed-length, locally significant identifier that is associated with a **forwarding equivalence class** (FEC), which represents a group of packets that are treated and forwarded in the same way through the network. The FEC groups packets that share the same forwarding path through the network, often based on factors such as their destination IP address, but the label itself does not represent or encode this address. Instead, the label is simply used to direct packets along the pre-determined path without referencing the full destination IP address at each routing step.

- **Label distribution**: MPLS routers use label distribution protocols, such as **Label Distribution Protocol (LDP)**, **Resource Reservation Protocol with Traffic Engineering** (RSVP-TE), or **Border Gateway Protocol (BGP)**, to exchange label information among themselves. Labels are distributed throughout the MPLS network, and routers build forwarding tables based on this information.

- **Label switching**: Instead of traditional IP routing, where routers make forwarding decisions based on the destination IP address, MPLS routers make forwarding decisions based on the MPLS label. When a packet enters the MPLS network, the ingress router attaches a label to the packet based on its FEC. As the packet traverses the network, intermediate routers forward it by performing label operations—namely, push, swap, and pop.

  - **Push**: When a packet enters the MPLS network, a label may be added (or "pushed") onto the packet.

  - **Swap**: When the packet reaches an intermediate router, the existing label is replaced (or "swapped") with a new one, which determines the next hop in the routing process.

    This operation is necessary because a label is a local construct on each router; they are dynamically allocated and not globally unique in traditional MPLS label distribution methods. Consequently, labels contain locally significant information for the routers. During the swap process, the router replaces the incoming label, which it had locally allocated and distributed, with the outgoing label received from the next-hop router along the path, ensuring efficient and accurate packet forwarding.

  - **Pop**: Finally, when the packet exits the MPLS network at the egress router, the label is removed (or "popped"), allowing the packet to be forwarded based on its original IP header.

    Additionally, **Penultimate Hop Popping (PHP)** may occur, where the penultimate router (the one just before the egress router) removes the label instead. This approach reduces processing at the egress router, allowing it to forward the packet directly using the original IP header, thereby enhancing overall network efficiency.

This label-based approach allows MPLS routers to make efficient forwarding decisions without needing to examine the IP header.

- **Label stack**: MPLS allows for the stacking of labels, forming a label stack. This is useful for scenarios such as MPLS **virtual private networks** (**VPNs**) and traffic engineering. Each label in the stack represents a different forwarding decision along the packet's path through the network. The top label, referred to as the transport label, is responsible for forwarding decisions within the MPLS network. In a multi-domain MPLS environment, multiple labels in the stack may serve as transport labels across different domain boundaries. Beneath these transport labels, there can be service labels, which represent L2-VPN or L3-VPN services. These service labels remain untouched throughout the packet's transit within the MPLS core and are only processed by the edge routers.

- **Traffic engineering**: MPLS supports traffic engineering, allowing network operators to optimize the use of network resources and control the flow of traffic through the network. By manipulating the labels and paths, operators can achieve better load balancing and resource utilization. It provides operators with greater flexibility to address the requirements of various use cases effectively.

- **Service agnostic operation**: One of the key benefits of MPLS features a BGP-free core, which reduces complexity by avoiding the need for core routers to carry extensive BGP routing tables. Additionally, mid-point core routers in MPLS networks operate without detailed service-specific forwarding information, showcasing inherent flexibility.

- **VPNs**: MPLS is commonly used for building Layer 2 and Layer 3 VPNs. Labels are used to distinguish between different VPNs, enabling the coexistence of multiple virtual networks over a shared physical infrastructure.

In essence, MPLS operates within the TCP/IP model, nestled between the traditional Layer 2 and Layer 3. This strategic placement allows MPLS to streamline packet forwarding through the network by utilizing a specialized header structure and MPLS operations.

Applications of MPLS

MPLS has found widespread adoption across various sectors and network types. It is frequently used in the following:

- **Internet service providers (ISPs)**: MPLS is essential for ISPs, enabling flexible, scalable services with multi-protocol capabilities and efficient traffic engineering, ensuring optimal network performance for customers.

- **VPNs**: MPLS-based VPNs provide secure, isolated communication channels over shared network infrastructure, offering businesses a cost-effective and secure way to interconnect their geographically dispersed offices.

- **Mobile networks**: The utilization of MPLS in mobile backhaul networks is on the rise, effectively managing the surge in traffic from the increasing number of mobile devices. This ensures a secure and smooth user experience, with the ability to separate internet, voice, signaling, and management traffic over a shared infrastructure.

- **Enterprise networks**: Enterprises benefit from MPLS by establishing secure and streamlined **wide area networks** (**WANs**) for various services. This enables reliable, high-speed data transmission across multiple locations, ensuring scalability, flexibility, and optimized traffic engineering for improved network performance.

MPLS is a versatile solution across diverse sectors and network landscapes, playing pivotal roles in enterprise networks, ISPs, VPNs, and mobile networks, each benefiting from its unique capabilities and advantages.

With a thorough understanding of MPLS and its benefits in enhancing network efficiency and performance, the focus now shifts to FRR, which serves as a critical mechanism within MPLS networks, providing rapid failure recovery to ensure minimal disruption in data transmission. This capability is crucial for maintaining high availability and reliability in modern networks. The upcoming section will delve into the principles of FRR.

Introducing FRR

FRR is a mechanism used in IP/MPLS networks to minimize the impact of network failures on the traffic flow. When a fault occurs in the network, the response varies depending on the protocol in use. Generally, there are two primary approaches to address the issue:

- **Wait for convergence**: In both traditional MPLS protocols, LDP and RSVP-TE, the network initiates a reconvergence process to identify the optimal new path from the source to the destination. The MPLS label distribution protocols rely on this reconvergence to re-establish the end-to-end MPLS path over the newly identified route. This period is known as **convergence delay**, during which traffic may be dropped, leading to the undesirable phenomenon of traffic black holes.

- **Local fast rerouting at the Point of Local Repair (PLR)**: While the network is reconverging, the router that detects the fault can reroute traffic around the failure, ensuring continuity of service. This diversion typically occurs within approximately 50 milliseconds, utilizing a precomputed backup path available on the PLR, referred to as the FRR path. Typically, all routers in the network are configured to support FRR, and the designation of the PLR may shift among routers based on the location of each fault. The FRR backup path remains active and carries traffic until the convergence process is complete and the new primary path is integrated into the routing and forwarding tables of the routers.

Although RSVP-TE protocols are outside the scope of this book, the LDP protocol is discussed in the next chapter, focusing primarily on MPLS forwarding rather than on failure and FRR mechanisms. The FRR method utilizing SR in this book is called **Topology Independent Loop-Free Alternate** (**TI-LFA**) and will be explored in greater detail in later chapters.

By promptly diverting traffic along pre-established alternate paths, FRR aims to minimize downtime and packet loss. This proactive approach ensures a resilient network infrastructure, allowing data packets to seamlessly reach their destinations despite potential disruptions in the network.

Here's a general understanding of how FRR works:

- **Protection paths**: FRR involves precomputing backup or protection paths that can quickly take over in the event of a failure. These paths are computed in advance and installed in the forwarding database for prompt activation when necessary.

- **Node or link failure detection**: When a failure occurs, such as a link or a node going down, the network devices detect the failure. This detection can be done through various mechanisms such as monitoring link status, using routing protocols, or employing dedicated protocols for fault detection, such as **Bidirectional Forwarding Detection (BFD)**.

- **PLR**: FRR commonly utilizes local repair mechanisms, where the affected node, upon detecting a failure, independently decides to reroute traffic through precomputed protection paths. This is a faster alternative to waiting for global network convergence. The node executing local repair is commonly referred to as the PLR.

- **FRR activation**: Upon detecting a failure, the network devices (routers or switches) quickly switch traffic to the pre-computed protection paths. This rapid switchover minimizes the impact on ongoing communications and helps maintain network connectivity.

FRR presents a proactive approach to minimize the effects of network failures. This is achieved by preemptively calculating and installing alternative routes for potential points of failure within the network. FRR serves as a robust defense against network downtime and packet loss. By ensuring uninterrupted traffic flow and reducing disruptions to ongoing communications and data in transit, FRR enhances the reliability and resilience of IP/MPLS networks. FRR does come with a cost, as it requires additional processing power and memory to compute and maintain backup path information in an immediately available state.

Upon acquiring a solid understanding of MPLS and establishing foundational knowledge of FRR, the stage is set to introduce SR in the upcoming section. The labs in this book focus on leveraging the best of both worlds, SR-MPLS and TI-LFA FRR. The synergy between the two helps create an enriching learning experience.

Introduction of Segment Routing (SR)

The fundamental distinction between traditional MPLS and SR lies in their simplicity. While traditional MPLS offers benefits such as traffic engineering, **Quality of Service (QoS)**, and support for L2 and L3 VPNs, SR architecture surpasses it. It works on the principle of source routing, where the state lies in the packet and a node routes a packet through an ordered list of instructions called *segments* embedded in the packet itself. It removes the need for complex signaling protocols and path states in the network.

What is SR?

Segment Routing (SR), or **Source Packet Routing in Networking (SPRING)**, is a routing technology that fundamentally transforms the routing framework, allowing the source node to determine the entire path of a packet, including all intermediate hops, from its origin to the final destination. This predetermined path is expressed as an ordered list of segments, stacked on the packet header. A segment, often identified by its **segment identifier (SID)**, may hold significance either locally to an SR node or globally within an SR domain.

At each hop along the predetermined path, the intermediate node determines its next hop by extracting information encoded in the packet header. This information, originally encoded by the upstream neighbor for the relevant segment, allows intermediate nodes to follow the segment sequence without the need for additional state information. This encoding streamlines the mechanism, reducing the overhead and complexity of traditional IP/MPLS protocols. By adopting SR, the source node gains enhanced control over routing decisions, providing a more adaptable and efficient approach to navigating complex network topologies.

One of the key advantages of SR is that it effectively reduces the network's statefulness. Since the entire path is encoded in the packet header, the network nodes don't need to store extensive routing information, as the segment or the list of segments itself carries all the necessary instructions. This stateless nature not only streamlines the network's operation but also enhances its scalability and resilience. Additionally, the reduced reliance on network-wide state information contributes to faster convergence during link or node failures. With its ability to provide efficient, flexible, and stateless routing, SR has emerged as a promising approach to tackle the challenges posed by modern networking environments.

SR is a flexible networking concept that is not tied to a specific data-plane technology. The architectural principles are detailed in RFC 8402 under the title *Segment Routing Architecture*. Notably, there are two primary implementations of SR:

- **SR-MPLS**: SR aligns with the original MPLS framework, which did not mandate the use of a specific signaling protocol. Consequently, SR utilizes various control-plane protocols, such as link-state routing protocols, to distribute or advertise segment information known as SIDs. Furthermore, no adjustments are required for the MPLS forwarding plane, as SIDs are encoded as *labels*, and a list of SIDs is represented as a *label stack*. These segments, depicted as labels, are distributed through protocols such as **Open Shortest Path First (OSPF)** or **Intermediate System to Intermediate System (IS-IS)** as IGP segments or via BGP as BGP segments.

- **SRv6**: The second implementation is SRv6, where SR leverages the IPv6 data plane. However, details about SRv6 are not covered in this particular book, as the IPv6 data plane operates quite differently from the MPLS data plane.

> **Note**
>
> The book exclusively focuses on SR-MPLS and its transition from the LDP, with IS-IS chosen as the **Interior Gateway Protocol (IGP)** for SR-MPLS label distribution. In the context of this book, the terms SR-MPLS and SR-IS-IS are used interchangeably, highlighting IS-IS as the selected IGP for establishing the SR-MPLS network.

MPLS has been a long-standing technology in networking, continuously evolving with the addition of various features over the years. One of the challenges with maintaining label integrity in MPLS lies in the traditional label distribution protocols, such as LDP and RSVP-TE. These protocols necessitate nodes to create and maintain sessions or states, leading to signaling message overheads that may demand extra computational power on routers. Additionally, maintaining signaling sessions and state information on routers can pose scalability challenges.

However, the next phase in its evolution is SR-MPLS.

Advantages of SR-MPLS

SR addresses several key problems in traditional MPLS networking, making it a powerful and flexible solution for various use cases.

Some of the significant problems that SR helps to solve include the following:

- **Label distribution**: Traditional MPLS uses LDP, RSVP-TE, or BGP for distributing labels whereas SR introduces label distribution through only IGP and BGP, allowing for greater flexibility and simplified label management.

- **Stateless operation**: LDP and RSVP-TE require routers to maintain state information about label bindings, leading to increased protocol complexity and memory usage. SR-MPLS operates in a stateless manner. It relies on source-based forwarding instructions encoded in the MPLS header, simplifying the control plane, ensuring efficient scalability, and providing flexibility for network changes.

- **Traffic engineering**: Built-in support for **SR-Traffic Engineering** (**SR-TE**), allowing operators to direct traffic through specific paths for optimized network performance.

Transitioning from traditional MPLS to SR-MPLS is a smooth process without any challenges, as described in the following steps:

- **Seamless network migration**: The label distribution in SR-MPLS differs from that of traditional protocols; however, the principles of label operations and forwarding adhere to the MPLS architecture. SR fits smoothly into current networks, making it easy to move from traditional MPLS networks to SR-MPLS networks.

- **Hardware support**: SR-MPLS utilizes the existing hardware-based data forwarding support inherent in MPLS. By building upon the foundation of the established MPLS infrastructure, SR-MPLS efficiently employs the same hardware components designed for MPLS data forwarding. It eliminates the necessity for significant hardware upgrades. It ensures a simple transition, reducing costs and operational challenges for organizations. This method, without causing disruptions, makes it easier for organizations to embrace new technologies.

Moving from traditional MPLS to SR-MPLS must be carefully planned, as the label distribution operation of the two protocols differs.

Disadvantages of SR-MPLS

SR provides a flexible and scalable way to steer traffic within a network. While it offers several advantages, there are also some potential disadvantages associated with SR. It's important to note that the impact of these disadvantages can vary depending on the specific use case and implementation. Here are some potential disadvantages:

- **Global segment allocation**: In traditional MPLS, label allocation is dynamic and automated, while SR-MPLS requires manual configuration of global segments (SIDs) as labels. Configuration errors in SR-MPLS can lead to routing issues, making careful planning and validation crucial.

- **Label stack depth limitations**: There are challenges associated with the label stack when too many labels are involved. Existing hardware may have limitations on the maximum depth of the MPLS label stack. SR-TE involves stacking labels to represent segments and enforce explicit paths, and in complex networks, the label stack can grow significantly. If the hardware cannot support a sufficient number of labels in the stack, it can limit the complexity and flexibility of the SR-TE paths.

To mitigate these potential drawbacks, network engineers need to carefully plan and validate the configuration of global segments (SIDs) to avoid routing issues. Additionally, it's essential to consider the hardware capabilities and limitations regarding label stack depth when designing SR-TE paths. Despite these challenges, with proper implementation and management, SR-MPLS can still offer significant benefits in terms of network flexibility and scalability. The upcoming section provides an insight into the structure of the book, with a brief description provided for each lab.

Structure of the book

The book is designed in such a manner that it takes you from a traditional MPLS network to an SR-MPLS network by introducing relevant features with theory as well as practical labs:

- *Lab 1, Getting Started with LDP-Based MPLS Network*, details the initial steps for you, explaining the IP schema, configuring IGP, and setting up traditional MPLS. This lab serves as the foundation for the subsequent chapters.

- *Lab 2, Introducing Segment Routing MPLS (SR-MPLS)*, breaks down the components of SR from both control and data plane perspectives, providing a comprehensive understanding of its building blocks.

- *Lab 3, SR-LDP Interworking*, explores the coexistence of SR and LDP as separate domains within the same IGP backbone. It covers forwarding traffic between SR and LDP domains.

- *Lab 4, Introducing TI-LFA (Topology Independent - Loop-Free Alternate)*, introduces the fast-reroute backup path mechanism, offering a means to protect SR-MPLS traffic in the core network.

- *Lab 5, Zero-Segment FRR*, demonstrates a backup path scenario without any additional segment, providing insights into specific configurations and scenarios.

- *Lab 6, Single-Segment FRR*, explores a backup path scenario with one additional segment.

- *Lab 7, Double-Segment FRR*, examines a backup path scenario with two additional segments.

- *Lab 8, Microloop Avoidance*, explores microloops avoidance explicit path scenario during topology changes.

- *Lab 9, TI-LFA Node Protection*, examines the TI-LFA node protection scenario, looking at backup path segments, and understanding how the calculation works.

- *Lab 10, TI-LFA Local SRLG-Disjoint Protection*, examines the TI-LFA local **Shared Risk Link Group (SRLG)**-disjoint protection scenario, which focuses on SRLG localized to the PLR, observing backup path components, and understanding the calculation process.

- *Lab 11, TI-LFA Global Weighted SRLG Protection*, examines the TI-LFA Global Weighted SRLG protection scenario, which includes remote SRLGs within the network rather than being localized to the PLR, observing backup path components, and understanding the calculation process.

- *Lab 12, TI-LFA Node + SRLG Protection*, examines the TI-LFA scenario that offers both node and SRLG protection simultaneously, observing backup path components, and understanding the calculation process.

- *Lab 13, TI-LFA Tiebreaker*, prepares the network topology for future TI-LFA tiebreaker scenarios, wherein multiple backup paths are available but not concurrently. It examines the TI-LFA scenario, preferring link protection over other options, with multiple backup paths available but not simultaneously; the TI-LFA SRLG protection scenario, where preference is given to SRLG protection, and multiple backup paths are available but not concurrently; and the preferred TI-LFA node protection scenario with multiple available backup paths, though not simultaneously.

The labs are carefully structured to be followed in a specific sequence, guiding users through the transition from traditional MPLS networks, which rely on LDP, to the full integration of SR-MPLS. Initially, they provide an in-depth understanding of how traditional MPLS networks operate, before gradually introducing the concepts of SR-MPLS. Eventually, the labs lead to a complete migration to SR-MPLS. Strategically placed within this sequence, the *TI-LFA* section enables you to grasp the significance of segments – in the case of SR-MPLS, labels – and their integration into packets. This section illuminates how these elements are incorporated into packets without the need for additional complex signaling protocols to maintain path state throughout the network, facilitating MPLS forwarding.

Network infrastructure

At its core, a **network infrastructure** comprises a combination of physical and virtual components that possess the ability to establish connectivity among themselves, facilitating the seamless flow of traffic. These elements include routers, switches, servers, and more. The size and complexity of the network infrastructure depend on the network's requirements and the scale of its operations.

In the lab described in this book, *building blocks* refer to the important parts or tools needed to set up the network system for education or learning. These building blocks are like the basic building materials that form the base of the lab, including specific software and hardware components.

The lab's building blocks, as described in the book, include the following:

- **Network operating system**: For the creation of network topologies in this book, the chosen network operating system is **Cisco IOS XR**, specifically using the **XRv9k** virtual image, a commercial product from Cisco (`https://www.cisco.com/`). While there are other paid and free network operating systems available, their compatibility with the labs presented in this book may vary, and they may not support all features or implementations in the same way.

  The Cisco XRv9k is a virtual router designed to run on x86-based servers. It emulates the Cisco IOS XR operating system, commonly used in service provider networks and high-end enterprise setups. To use the Cisco XRv9k, a service contract with Cisco is required, which can be obtained by contacting a Cisco account manager.

  For the labs conducted in this book, the Cisco XRv9k version 7.5.2 image was used and configured with 4 vCPUs, 16384 MB RAM, and 11 Ethernet ports to meet the lab's specific requirements. The vCPU and RAM are configured following Cisco's guidelines as stated on their official website.

  Cisco also offers an IOS-XR-based container image called **Cisco XRd**, but testing revealed that microloops avoidance (covered later in the book) did not function as expected on it.

- **Network emulator**: The labs in this book have been tested on two network emulation platforms: EVE-NG (both the Professional and Community editions) and Containerlab.

  - **EVE-NG**: EVE-NG (`https://www.eve-ng.net/`) is a network emulation platform that allows users to create virtual network topologies through a web-based drag-and-drop interface. It provides networking professionals with the capability to design, test, and troubleshoot network configurations in a secure and virtualized environment. The labs in this book, along with the configurations and outputs, are demonstrated using **EVE-NG Professional** (the paid version). However, they have also been successfully tested on the free **EVE-NG Community Edition**, ensuring compatibility with both versions.

  - **Containerlab**: Containerlab (`https://containerlab.dev/`) enables the creation of virtual network topologies using containers of network operating systems, connecting them according to a `.yml` topology definition file. The Cisco XRv9k 7.5.2 image was successfully containerized following the instructions provided on the official website for use with Containerlab. The labs were then tested and validated against the network topology using XRv9k virtual machines within containers.

The EVE-NG Community Edition and Containerlab are both free to use, so you can choose either based on your preference.

> **Note**
>
> There are other tools that can also be successfully used to run the labs in this book, such as **Cisco Modeling Labs** (**CML**), which requires a purchased license from Cisco, and **GNS3**, a free platform for simulating network topologies.

- **Server**: EVE-NG and Containerlab can be installed on either a physical server located on-site or a cloud platform. Comprehensive installation guides and instructions for onboarding virtual router images are available on their respective websites. In this book's labs, a robust physical server with 96 vCPU, 1 TB RAM, and 2 TB SSD was utilized. This powerful configuration ensured the smooth execution of complex network simulations and virtual machines, providing an engaging and effective learning experience.

> **Google Cloud Platform (GCP)**
>
> Those interested in hosting it on GCP can take advantage of a $300/90-day free trial for new users. This trial period allows users to leverage cloud resources to host virtual machines and perform network simulations in a scalable and flexible environment.

This lab has also been successfully tested on GCP using EVE-NG Community Edition and Containerlab, each deployed separately on different virtual machines.

For running the lab on GCP with EVE-NG or Containerlab, it is recommended to choose the `n2-standard-32` machine type from the N2 machine types within the general-purpose family. This configuration provides 32 vCPUs and 128 GB of RAM. Additionally, select a 100 GB SSD for storage.

When combined, these building blocks create a powerful lab environment for learning and practicing networking concepts. The personal server setup utilizes EVE-NG Pro as the platform to design and visualize network topologies, while the Cisco XRv9k provides the virtual router necessary for running Cisco IOS XR-based configurations.

It's worth noting that the specifics of the building blocks may vary based on the context and publication date. As technology constantly evolves, it is essential to refer to the latest information and official documentation from the respective providers for accurate details and setup instructions.

You can find the GitHub repository for this book at the following link: `https://github.com/PacktPublishing/Segment-Routing-in-MPLS-Networks`.

The topology definition files for EVE-NG, ContainerLab, and CML are as follows:

- **EVE-NG**: `https://github.com/PacktPublishing/Segment-Routing-in-MPLS-Networks/blob/main/Segment_Routing_MPLS.zip`

- **Containerlab**: `https://github.com/PacktPublishing/Segment-Routing-in-MPLS-Networks/blob/clab-srmpls/srmpls-clab-topo.yml`

- **CML**: https://github.com/PacktPublishing/Segment-Routing-in-MPLS-Networks/blob/main/CML_SRMPLS.yaml

You can import the topology definition file for your preferred platform to test the labs alongside the book.

> **Note**
>
> The labs are curated in a specific sequence. If a particular scenario from the middle chapters needs to be explored, the book's GitHub repository includes separate Ansible playbooks for EVE-NG and Containerlab to automate the prerequisite labs.

The EVE-NG ansible playbook for the book is available at the following URL: https://github.com/PacktPublishing/Segment-Routing-in-MPLS-Networks/tree/ansible-srmpls.

The Containerlab Ansible playbook for the book is available at the following URL: https://github.com/PacktPublishing/Segment-Routing-in-MPLS-Networks/tree/clab-srmpls.

The main repository also includes the startup configurations and the specific configurations used for each chapter of the book.

Topology

Designing a topology in EVE-NG or Containerlab is a relatively straightforward process, but for those who are new to the platform, there is ample documentation available on their website to guide you through the process effectively. The documentation provides users with comprehensive instructions and resources to create, configure, and manage network topologies effortlessly. So, even if you're just starting, you can easily leverage the documentation to make the most of this powerful network emulation tool and set up your desired network scenarios with ease.

The following information describes the topology that will be used as part of the transition from traditional MPLS to SR-MPLS in the book. The topology consists of several nodes interconnected through point-to-point links.

It is recommended that you replicate the network topology outlined in this book. Notably, an additional router, designated as P9, will be introduced later in the book. This information is aimed at ensuring a clear understanding of the hardware requirements essential for establishing this virtual setup.

The network setup outlined in this book spans across all the labs, each intended to be followed in a sequential manner. In a standard MPLS network, both a source edge router and a destination edge router are typically utilized. However, for the FRR labs, an additional component known as a PLR is necessary.

To ensure clarity and ease of understanding in visualizing packet flow, the following topology diagram has been created with a minimal number of routers:

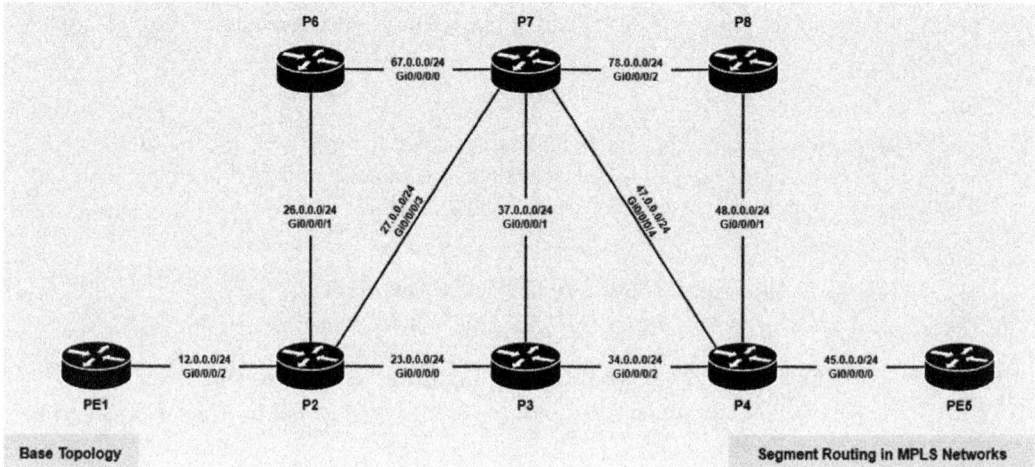

Figure 1.2 – Base topology

In this setup, PE1 serves as the source node while *PE5* functions as the destination. P2, given its multiple connections within the core network, is utilized as the PLR node for TI-LFA labs, offering alternative paths for forwarding traffic in the event of potential failures.

> **Note**
> The words *router*, *node*, or *device* will be used interchangeably in this book.

The book describes a base network using the Cisco IOS-XR network operating system, incorporating the IS-IS IGP routing protocol and LDP as the traditional MPLS label distribution protocol. The initial lab will focus on configuring and operating this foundational topology, with subsequent labs in the book building upon these fundamentals. The combination of these protocols forms a solid groundwork for building and understanding SR-MPLS forwarding.

IP address schema

The book follows a particular IP address scheme. Having a clearly defined addressing plan is crucial to enable effective communication among devices within the network.

The following elaborates on the schema applied to simplify the understanding of the node names and their corresponding IP addresses:

- **Nodes/routers**: The topology comprises multiple nodes, each identified by a label. PE1 and P2 are two of the nodes described, and they are used to clarify the terminology in the next sections.

- **Loopback IP addresses**: A loopback address is a special type of IP address used to test network connectivity on a local device without actually sending data over the physical network. The loopback interface is virtual and does not represent any specific physical network interface on the device. It is often used for diagnostic purposes, testing applications, and local communication. The loopback interface is commonly used as the router ID by various protocols, defining the router's identity. Unlike physical connections, the loopback remains operational as long as the router itself is running.

  The nomenclature for loopback addresses is as follows: x.x.x.x/32, where x is the node number.

  Here are some examples:

  - For PE1, the loopback address is 1.1.1.1/32
  - For P2, the loopback address is 2.2.2.2/32, and so on

- **Point-to-point link IP addresses**: The topology includes point-to-point links connecting different nodes. Each link has its IP addresses assigned to the interfaces on either end of the link.

  The nomenclature for point-to-point interface addresses is xy.0.0.y/24, where x < y.

  For example, the point-to-point link between PE1 and P2 gets assigned the IP address 12.0.0.1/24 on PE1 and 12.0.0.2/24 on P2.

  Here, x = 1 and y = 2, satisfying the condition x < y.

Overall, this topology illustrates a network configuration with specific IP addressing for loopback, and point-to-point links.

IGP

Within the context of this book, the **IGP** of choice is IS-IS. The network architecture involves routers interconnected in a flat, contiguous **Level 2 (L2)** backbone. To enhance simplicity and ease of computation, a standardized metric of 10 has been uniformly assigned to all router connections, unless specific variations are indicated. This metric system provides a straightforward means of estimating path cost, aligning directly with the number of hops between routers. The entire network operates within the IS-IS L2 domain, ensuring seamless participation of all routers in this routing protocol.

LDP

In this book, LDP is employed as the traditional label distribution protocol for the starting topology and the first lab, initiating the guide on SR-MPLS. This decision is influenced by its minimal configuration needs, and both LDP and SR-MPLS seamlessly integrate with the IGP, staying aligned with the book's focus without unnecessary diversions.

> **Note**
>
> The sequence of labs is carefully curated to minimize the need for configuration changes as you progress from one lab to the next. This approach is intended to enhance focus on understanding different scenarios while directing attention specifically to the relevant pieces of configuration.

Summary

By now, you will have acquired a comprehensive grasp of MPLS and its pivotal role in modern networking. Fundamental MPLS concepts, spanning its definition, applications, and key features such as FRR and SR, have been thoroughly explored.

MPLS emerges as a versatile tool, catering to diverse networking needs, including efficient traffic management, enhanced QoS, and robust VPN support.

SR represents a paradigm shift in routing architectures, offering streamlined network operations and improved scalability while maintaining existing data-plane forwarding mechanisms. SR-MPLS seamlessly integrates SR with traditional MPLS networks, facilitating eventual migration to SR-only MPLS environments without major disruption.

FRR stands out as a critical element in enhancing network resilience, proactively establishing alternative paths that can swiftly activate in response to network failures.

The chapter has also outlined the network infrastructure tailored for the book's purpose, emphasizing simplicity to aid your comprehension. The topology and structure of the book have been meticulously crafted to guide you seamlessly through topics, ensuring a fluid learning experience conducive to skill development.

In the next chapter, the foundational lab will utilize IS-IS as the IGP and LDP as the traditional MPLS protocol. These choices reflect their prevalence and effectiveness within service provider networks, providing you with practical insights into real-world deployments.

References

The following are the references used in this chapter, offering additional resources and further reading to deepen your understanding of the concepts discussed:

- `https://www.segment-routing.net/`
- `https://www.cisco.com`
- `https://containerlab.dev/`
- `https://www.eve-ng.net/`
- `https://cloud.google.com/`
- *Segment Routing Part I*, by Clarence Filsfils, Kris Michielsen, Ketan Talaulikar
- *[RFC1180] Socolofsky, T. and C. Kale, "TCP/IP tutorial", RFC 1180, January 1991,* <`http://www.rfc-editor.org/info/rfc1180`>.

Lab 1 – Getting Started with LDP-Based MPLS Network

Welcome to the first hands-on lab of this book, which focuses on establishing a foundational topology utilizing the traditional MPLS protocol LDP. This lab starts from scratch, beginning with the creation of a base physical topology in a virtual environment. It then explains and configures the IP address conventions used for all routers and the links connecting them, facilitating ease of understanding and identification. The IS-IS link-state IGP routing protocol is deployed for end-to-end IP communication, while LDP is used for end-to-end MPLS communication between all the routers in the topology.

This lab is more extensive than others in the book, as it forms the basis for subsequent chapters. The IS-IS routing protocol will later carry SR-MPLS control-plane information, and MPLS will provide data-plane forwarding for network traffic. Therefore, significant emphasis is placed on understanding how IS-IS carries link-state information and how MPLS enables labeled forwarding. This comprehensive explanation ensures a solid grasp of both protocols, making the remaining labs easier to absorb.

The following section outlines the topics covered in the book, presenting them as objectives for the lab.

Objective

The primary goal of the initial lab is to set up a fundamental MPLS network based on LDP. The key emphasis lies in verifying the functionality of traditional MPLS forwarding across all routers in the topology. The end-to-end reachability test will specifically concentrate on the IP and label forwarding from PE1 to PE5, as shown in the following topology.

This single topology will be utilized throughout the book for all hands-on labs.

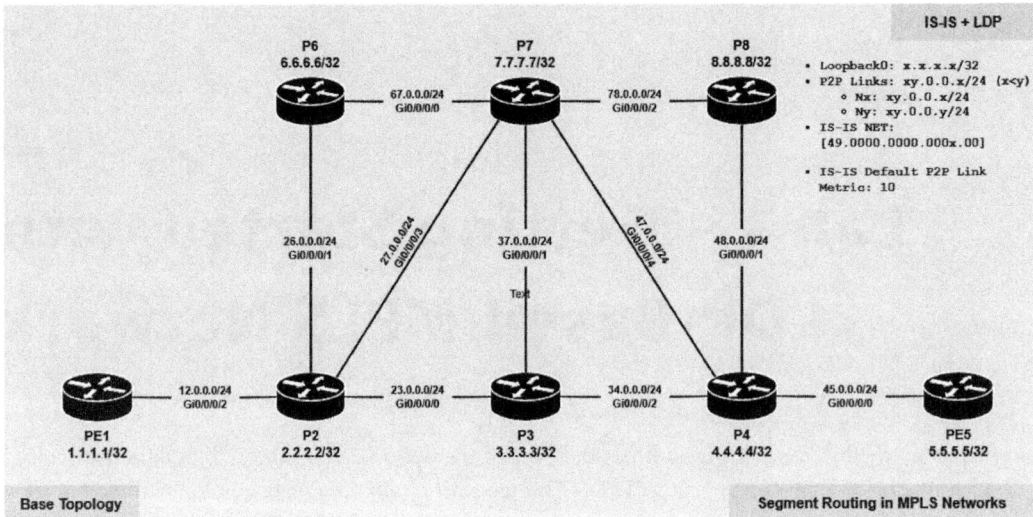

Figure 2.1 – Base topology

To achieve the goal, a three-step process will be followed:

- **IPv4 address**

  I.   Develop a simple yet efficient IP addressing schema and configure the routers accordingly. This makes it easier to identify both routers (loopback) and the connections between them.

  II. Verify that all links are correctly configured by checking IP reachability on the point-to-point connections between the routers.

- **IGP – IS-IS (Intermediate System - Intermediate System)**

  I.   Configure IS-IS (Intermediate System - Intermediate System) as the **Interior Gateway Protocol (IGP)**.

  II. Verify IS-IS adjacency by confirming that routers establish the required connections with each other on directly connected point-to-point links.

  III. Explore the IS-IS database and analyze IS-IS link state packets to gain a more thorough understanding of the network's topology.

  IV. Verify end-to-end IP reachability among all routers to ensure IGP routing functions as expected.

- **MPLS – LDP (Label Distribution Protocol)**

    I.   Configure the **Label Distribution Protocol (LDP)** for label distribution throughout the network.

    II.   Verify LDP discovery and session establishment among routers.

    III. Verify MPLS **Label Information Base (LIB)** for insights into labels learned from downstream LDP neighbor routers.

    IV.   Verify MPLS **Label Forwarding Information Base (LFIB)** to confirm the labels installed in the data plane.

    V.   Verify **Forwarding Information Base (FIB)** to confirm the labels installed in the data plane for IP-to-MPLS traffic.

    VI.   Verify end-to-end MPLS reachability among all routers to ensure label forwarding functions as expected.

> **Note**
> Almost all of the outputs checked in all the labs are from the perspective of the source router PE1 or P2 to the destination router PE5.

The first step begins with the following section.

IPv4 address

The IP addressing schema selected for this book follows the template outlined here:

- **Loopback Interface IP**: x.x.x.x/32, where x is the router number, 1 for PE1, 2 for P2 and so on.

- **Physical Interface IP**: The **Point-to-Point (P2P)** link subnet is xy.0.0.0/24, where x < y

    Nx: xy.0.0.x/24, Ny: xy.0.0.y/24, for link between Nx: PE1 and Ny: P2, it would be 12.0.0.1/24 and 12.0.0.2/24, respectively, and so on

The loopback IP schema helps identify the router easily, while the P2P link IP schema helps in identifying the routers at either end of a particular link. The next section provides the template using this schema.

Template

The template used for configuring the loopback interface is as follows.

Interface Loopback0

The best practice for IP address configuration of a loopback interface is to use a host address with a /32 notation, as it does not participate in any subnet, as shown here:

```
interface Loopback0
 ipv4 address {{ x.x.x.x }} 255.255.255.255
```

The preceding commands are explained next:

- `interface Loopback0`: This command enters configuration mode for loopback interface 0
- `ipv4 address {{ x.x.x.x }} 255.255.255.255`: This sets the IPv4 address for Loopback0 to x.x.x.x with a subnet mask of 255.255.255.255 equivalent to /32 notation

A loopback interface turns operationally up by default after configuration, unless explicitly shut down.

> **Note**
>
> In MPLS networks, the loopback address must be a /32 to ensure that the label is distributed for specific IP addresses, allowing for accurate identification of the routers within the network.

The template used for configuring the physical interface is as follows.

Physical interface

The best practice for IP address configuration of a physical interface in a production network is to use a /31 subnet. However, for the network topology used in this book, a /24 subnet is chosen to allow router numbers to be used in the IP address for easier identification of the subnet and the routers. This fits the schema designed for the lab.

```
interface {{ interface_name }}
 description "P(E){{ x }}_to_P(E){{ y}}"
 ipv4 address {{ xy.0.0.(x or y) }} 255.255.255.0
 no shutdown
```

The preceding commands are explained here:

- `interface {{ interface_name }}`: This command enters configuration mode for a physical interface specified by `{{ interface_name }}`.
- `description "P(E){{ x }}_to_P(E){{ y}}"`: This adds a description to the interface, indicating that it connects from `"P(E){{ x }}_to_P(E){{ y}}"`.
- `ipv4 address {{ xy.0.0.(x or y) }} 255.255.255.0`: This sets the IPv4 address for the interface to xy.0.0.(x or y) with a subnet mask of 255.255.255.0, equivalent to /24 notation. The values x and y in the address will be replaced with specific numbers as per the IP address schema explained earlier.

- `no shutdown`: This command ensures that the interface is enabled and brought up, allowing it to actively participate in network communication.

Now that the commands and the template are understood, the routers are configured as follows, utilizing the schema and the template defined in the previous section.

Configuration

The following section illustrates the IP address configuration applied for each router for both their loopback and physical interfaces.

> **Note**
>
> On the Cisco IOS-XR platform, the `configure` command is required to enter the platform's global configuration mode, and the `commit` command is required to push the target configuration to the running configuration. After making changes, you can exit to the exec mode with the `exit` command.

The Cisco IOS-XRv9k image employs Gigabit Ethernet as a physical interface, which is the type of interface used to interconnect the routers in this topology. The schema, when applied to the values for each router and the interconnect, generates the configuration for each router as outlined here:

- The PE1 router is connected only to router P2. The schema for each interface is defined as follows:

    - For the `loopback0` interface:

      - PE1: The value of `x = 1`

    - For the following physical interface:

      - `PE1 - P2`: The value of `x = 1, y = 2 (where x < y)`

The IPv4 address template is applied using the specified values.

PE1

```
interface Loopback0
 ipv4 address 1.1.1.1 255.255.255.255

interface GigabitEthernet0/0/0/2
 description "PE1_to_P2"
 ipv4 address 12.0.0.1 255.255.255.0
 no shutdown
```

The preceding configuration is generated and applied accordingly.

- The P2 router is connected to the PE1, P3, P6, and P7 routers. The schema for each interface is defined as follows:

  - For the loopback0 interface:

    - P2: The value of x = 2

  - For the following physical interface:

    - P2 - P3: The value of x = 2, y = 3 (where x < y)
    - P2 - P6: The value of x = 2, y = 6 (where x < y)
    - P2 - PE1: The value of x = 1, y = 2 (where x < y)
    - P2 - P7: The value of x = 2, y = 7 (where x < y)

The IPv4 address template is applied using the specified values.

P2

```
interface Loopback0
 ipv4 address 2.2.2.2 255.255.255.255

interface GigabitEthernet0/0/0/0
 description "P2_to_P3"
 ipv4 address 23.0.0.2 255.255.255.0
 no shutdown

interface GigabitEthernet0/0/0/1
 description "P2_to_P6"
 ipv4 address 26.0.0.2 255.255.255.0
 no shutdown

interface GigabitEthernet0/0/0/2
 description "P2_to_PE1"
 ipv4 address 12.0.0.2 255.255.255.0
 no shutdown

interface GigabitEthernet0/0/0/3
 description "P2_to_P7"
 ipv4 address 27.0.0.2 255.255.255.0
 no shutdown
```

The preceding configuration is generated and applied accordingly.

- The P3 router is connected to routers P2, P7, and P4. The schema for each interface is defined as follows:

  - For the `loopback0` interface:

    - P3: The value of x = 3

  - For the following physical interface:

    - P3 - P2: The value of x = 2, y = 3 (where x < y)
    - P3 - P7: The value of x = 3, y = 7 (where x < y)
    - P3 - P4: The value of x = 3, y = 4 (where x < y)

  The IPv4 address template is applied using the specified values.

  P3

  ```
  interface Loopback0
    ipv4 address 3.3.3.3 255.255.255.255

  interface GigabitEthernet0/0/0/0
    description "P3_to_P2"
    ipv4 address 23.0.0.3 255.255.255.0
    no shutdown

  interface GigabitEthernet0/0/0/1
    description "P3_to_P7"
    ipv4 address 37.0.0.3 255.255.255.0
    no shutdown

  interface GigabitEthernet0/0/0/2
    description "P3_to_P4"
    ipv4 address 34.0.0.3 255.255.255.0
    no shutdown
  ```

  The preceding configuration is generated and applied accordingly.

- The P4 router is connected to the P3, P7, PE5, and P8 routers. The schema for each interface is defined as follows:

  - For the `loopback0` interface:

    - P4: The value of x = 4

  - For the following physical interface:

    - P4 - PE5: The value of x = 4, y = 5 (where x < y)
    - P4 - P8: The value of x = 4, y = 8 (where x < y)

- P4 - P3: The value of x = 3, y = 4 (where x < y)

- P4 - P7: The value of x = 4, y = 7 (where x < y)

The IPv4 address template is applied using the specified values.

P4

```
interface Loopback0
 ipv4 address 4.4.4.4 255.255.255.255

interface GigabitEthernet0/0/0/0
 description "P4_to_PE5"
 ipv4 address 45.0.0.4 255.255.255.0
 no shutdown

interface GigabitEthernet0/0/0/1
 description "P4_to_P8"
 ipv4 address 48.0.0.4 255.255.255.0
 no shutdown

interface GigabitEthernet0/0/0/2
 description "P4_to_P3"
 ipv4 address 34.0.0.4 255.255.255.0
 no shutdown

interface GigabitEthernet0/0/0/4
 description "P4_to_P7"
 ipv4 address 47.0.0.4 255.255.255.0
 no shutdown
```

The preceding configuration is generated and applied accordingly.

- The PE5 router is connected only to router P4. The schema for each interface is defined as follows:

  - For the loopback0 interface:

    - PE5: The value of x = 5

  - For the following physical interface:

    - PE5 - P4: The value of x = 4, y = 5 (where x < y)

The IPv4 address template is applied using the specified values.

PE5

```
interface Loopback0
 ipv4 address 5.5.5.5 255.255.255.255

interface GigabitEthernet0/0/0/0
```

```
  description "PE5_to_P4"
  ipv4 address 45.0.0.5 255.255.255.0
  no shutdown
```

The preceding configuration is generated and applied accordingly.

- The P6 router is connected to the P2 and P7 routers. The schema for each interface is defined as follows:

  - For the loopback0 interface:

    - P6: The value of x = 6

  - For the following physical interface:

    - P6 - P7: The value of x = 6, y = 7 (where x < y)

    - P6 - P2: The value of x = 2, y = 6 (where x < y)

The IPv4 address template is applied using the specified values.

P6

```
interface Loopback0
  ipv4 address 6.6.6.6 255.255.255.255

interface GigabitEthernet0/0/0/0
  description "P6_to_P7"
  ipv4 address 67.0.0.6 255.255.255.0
  no shutdown

interface GigabitEthernet0/0/0/1
  description "P6_to_P2"
  ipv4 address 26.0.0.6 255.255.255.0
  no shutdown
```

The preceding configuration is generated and applied accordingly.

- The P7 router is connected to the P2, P3, P4, P6, and P8 routers. The schema for each interface is defined as follows:

  - For the loopback0 interface:

    - P7: The value of x = 7

  - For the following physical interface:

    - P7 - P6: The value of x = 6, y = 7 (where x < y)

    - P7 - P3: The value of x = 3, y = 7 (where x < y)

- P7 - P8: The value of x = 7, y = 8 (where x < y)

- P7 - P2: The value of x = 2, y = 7 (where x < y)

- P7 - P4: The value of x = 4, y = 7 (where x < y)

The IPv4 address template is applied using the specified values.

P7

```
interface Loopback0
 ipv4 address 7.7.7.7 255.255.255.255

interface GigabitEthernet0/0/0/0
 description "P7_to_P6"
 ipv4 address 67.0.0.7 255.255.255.0
 no shutdown

interface GigabitEthernet0/0/0/1
 description "P7_to_P3"
 ipv4 address 37.0.0.7 255.255.255.0
 no shutdown

interface GigabitEthernet0/0/0/2
 description "P7_to_P8"
 ipv4 address 78.0.0.7 255.255.255.0
 no shutdown

interface GigabitEthernet0/0/0/3
 description "P7_to_P2"
 ipv4 address 27.0.0.7 255.255.255.0
 no shutdown

interface GigabitEthernet0/0/0/4
 description "P7_to_P4"
 ipv4 address 47.0.0.7 255.255.255.0
 no shutdown
```

The preceding configuration is generated and applied accordingly.

- The P8 router is connected to the P4 and P7 routers. The schema for each interface is defined as follows:

  - For the loopback0 interface:

    - P8: The value of x = 8

- For the following physical interface:

  - P8 - P4: The value of x = 4, y = 8 (where x < y)

  - P8 - P7: The value of x = 7, y = 8 (where x < y)

The IPv4 address template is applied using the specified values.

P8

```
interface Loopback0
 ipv4 address 8.8.8.8 255.255.255.255

interface GigabitEthernet0/0/0/1
 description "P8_to_P4"
 ipv4 address 48.0.0.8 255.255.255.0
 no shutdown

interface GigabitEthernet0/0/0/2
 description "P8_to_P7"
 ipv4 address 78.0.0.8 255.255.255.0
 no shutdown
```

The preceding configuration is generated and applied accordingly.

This completes the IP address configuration of all the routers in the topology and enables IP communication between them on their physical connections, which is verified in the next section.

Verification

The following sections verify that the configuration is applied to the interfaces and that there is IP communication on the point-to-point links.

Verifying interface IP address

The following output from the PE1 router confirms that the IP addresses are applied according to the configurations on each router. Additionally, the configured interfaces are operational and up and running:

```
RP/0/RP0/CPU0:PE1#show ip interface brief

Interface              IP-Address      Status        Protocol Vrf-Name
Loopback0                  1.1.1.1     Up            Up       default
MgmtEth0/RP0/CPU0/0        192.168.18.1 Up           Up       mgmt
GigabitEthernet0/0/0/0     unassigned  Down          Down     de-
fault
GigabitEthernet0/0/0/1     unassigned  Down          Down     de-
fault
```

GigabitEthernet0/0/0/2 fault	12.0.0.1	Up	Up	de-
GigabitEthernet0/0/0/3 fault	unassigned	Down	Down	de-
GigabitEthernet0/0/0/4 fault	unassigned	Down	Down	de-
GigabitEthernet0/0/0/5 fault	unassigned	Down	Down	de-
GigabitEthernet0/0/0/6 fault	unassigned	Down	Down	de-
GigabitEthernet0/0/0/7 fault	unassigned	Down	Down	de-
RP/0/RP0/CPU0:PE1#				

After configuring all the routers, they should be able to reach each other using the directly connected links, as confirmed in the next section.

Verifying P2P IP reachability

From router PE1, ping the far-end IP address of the link connected to router P2, 12.0.0.2. The successful ping test that follows verifies that IP reachability functions as expected on the local link between PE1 and P2.

The same outcome is expected on all other point-to-point core links in the topology.

```
RP/0/RP0/CPU0:PE1#ping 12.0.0.2
Type escape sequence to abort.
Sending 5, 100-byte ICMP Echos to 12.0.0.2, timeout is 2 seconds:
!!!!!
Success rate is 100 percent (5/5), round-trip min/avg/max = 2/2/3 ms
RP/0/RP0/CPU0:PE1#
```

The routers can reach each other only over the directly connected point-to-point links; routers that are not directly connected are unreachable as they do not have routes to remote subnets.

In the next section, the IS-IS link-state **Interior Gateway Protocol** (IGP) is configured to enable end-to-end routing among all the routers in the topology.

Interior Gateway Protocol – IS-IS (Intermediate System - Intermediate System)

IS-IS is an interior gateway link-state routing protocol used in large enterprise and service provider networks. It operates at the data link layer (Layer 2) and uses the **Link State Protocol Data Units** (**LSPDUs**) to exchange information between directly connected routers. The protocol is mainly utilized to build and maintain a consistent view of the network topology, allowing routers to make efficient routing decisions.

The original ISO IS-IS protocol was adapted to additionally carry IP prefixes and is known as Integrated IS-IS or Dual IS-IS, hereafter referred to as IS-IS.

Here's a summary of how IS-IS operates:

- **Neighbor Discovery**: When a router running IS-IS is initialized or connected to the network, it begins the process of discovering its neighboring routers. IS-IS uses a specialized `Hello` protocol to exchange `Hello` packets with other routers on the same broadcast or point-to-point network segment. Through this process, routers establish adjacencies with their neighbors.

- **Link State PDU (LSPDU) Generation**: Upon establishing adjacencies, each router generates a **Link State Protocol Data Unit** (**LSPDU**) containing information about itself and its directly connected links. This information encompasses the link's IP address, metrics (cost or weight), and other attributes.

- **LSPDU Flooding**: In this process, routers transmit LSPDUs containing their local network information to neighboring routers. These neighbors update their database and then flood these LSPs to their neighbors, ensuring a widespread distribution. This flooding mechanism guarantees that all routers maintain up-to-date knowledge of the entire network's structure. The goal of flooding is to ensure that, by the end of the process, each router has an identical LSDB copy.

- **Link State Database (LSDB) Construction**: Upon receiving LSPDUs from neighboring routers, each router constructs its **Link State Database** (**LSDB**), which contains the information from all the received LSPDUs. The LSDB serves as a local representation of the network's topology.

- **Shortest Path First (SPF) Algorithm**: With a complete LSDB, each router executes the SPF algorithm (also known as Dijkstra's algorithm) independently to calculate the shortest path to every other router in the network. The SPF algorithm uses the link metrics in the LSPDUs to determine the best path to each destination router.

- **Routing Table**: Based on the SPF calculation, each router populates its routing table, which stores best-path forwarding information about directly connected and remote destination networks, the next hop to reach them, and the interface to use. The routing table is populated by connected routes, static routes, and dynamic routing protocols such as IS-IS, OSPF, or BGP, and helps routers make efficient forwarding decisions.

This achieves the initial network convergence and establishes IP reachability among the IS-IS routers. The IS-IS protocol uses periodic and triggered updates, as outlined next, to maintain a real-time view of the network:

- **Periodic Updates**: IS-IS routers continue to exchange LSPDUs periodically to keep the network topology information up to date and synchronize their databases for accurate routing.

- **Triggered Updates**: IS-IS aims to achieve rapid convergence in the event of network changes. When a link or router failure occurs, the affected router immediately generates a new LSPDU reflecting the change and floods it to the entire network. Routers receiving this update quickly recalculate the SPF to update their routing tables and find alternate paths, if available, to reach the affected destination.

The preceding points illustrate how IS-IS ensures efficient and reliable routing through link-state information propagation throughout the network. The configuration required to enable IS-IS routing in the lab topology is explored next.

Template

This chapter utilizes the IS-IS protocol configuration template outlined here:

```
router isis IGP
is-type level-2-only
net 49.0000.0000.000{{ x }}.00
log adjacency changes
address-family ipv4 unicast
  metric-style wide Level-2

interface Loopback0
  passive
  address-family ipv4 unicast

interface {{ interface_name }}
  circuit-type level-2-only
  point-to-point
  hello-padding disable
  address-family ipv4 unicast
   metric 10
```

Now, let's break down the details.

Protocol

The commands applied under the protocol context define how it is set up on the router and also define its unique identity in the network. The commands used in this lab are explained here:

- `router isis IGP`: This command enables the IS-IS routing protocol and sets the process name to IGP.

- `is-type level-2-only`: This specifies that the router will only participate in Level-2 IS-IS routing. The routers in Level-1 areas manage intra-area routing, as they lack knowledge of other areas. Meanwhile, the routers in Level-2 areas handle inter-area routing and function as a backbone connecting multiple areas. The provider network routers are typically configured as Level-2 area routers.

- `net 49.0000.0000.000{{ x }}.00`: This is the **Network Entity Title** (**NET**) address assigned to the router. The NET address is used to uniquely identify routers in the IS-IS network. The `{{ x }}` placeholder will be replaced with the number of the node ID.

- `log adjacency changes`: This command enables logging for adjacency changes, such as going up or down.

IS-IS supports multi-topology such as the IPv4 and IPv6 address families, but for the purposes of this book, only the IPv4 address family is used. The template configuration for this is explained next. The address-family context configuration is as follows:

- `address-family ipv4 unicast`: This command configures the router for IPv4 unicast routing within the IS-IS process.
- `metric-style wide Level-2`: This sets the metric style for Level-2 routes to `wide`. IS-IS uses metrics to determine the best path for routing. Narrow metrics utilize a 6-bit field, allowing for a range of values from 0 to 63. In contrast, wide metrics employ a 24-bit field, enabling a range from 0 to 16,777,215.

The commands applied under the preceding protocol address-family context apply to all interface address-family contexts too, unless overridden by another one within the interface address-family context.

The specifics for the interfaces are covered next.

Loopback0 interface

The loopback interface configuration is as follows:

- `interface Loopback0`: This command enters configuration mode for loopback interface 0
- `passive`: This sets the loopback interface to passive mode, meaning it will not participate actively in routing updates but will still advertise its existence
- `address-family ipv4 unicast`: This command configures the loopback interface for IPv4 unicast routing within the IS-IS process

The loopback interface is a host interface. It is not connected to any other router, so it is configured as passive to prevent the router from sending updates over that interface and to conserve resources.

Physical interface

The physical interface configuration differs from the loopback because it is used to exchange IS-IS information with adjacent routers.

The template commands are explained here:

- `interface {{ interface_name }}`: This command enters configuration mode for a physical interface specified by `{{ interface_name }}`.
- `circuit-type level-2-only`: This specifies that the interface will only participate in Level-2 IS-IS routing. Since the routers are participating only in a Level-2 area, and due to the `is-type` configuration set to `level-2-only`, the `circuit-type` configuration is also set to `level-2-only` to maintain consistency. This alignment is configured because the routers cannot establish adjacency over L1 (Level-1) areas.

- `point-to-point`: This sets the interface to point-to-point mode, indicating it connects directly to another router and is not part of a broadcast domain. An ethernet circuit represents a broadcast domain, so it must be configured as point-to-point.

- `hello-padding disable`: This disables hello padding on the interface. Hello padding is used to increase the size of IS-IS Hellos to detect MTU mismatch-related failures more quickly, but it's disabled here. While the initial packets continue to be padded, subsequent packets do not undergo padding.

- `address-family ipv4 unicast`: This command configures the interface for IPv4 unicast routing within the IS-IS process.

- `metric 10`: This sets the metric for routes through this interface to 10. Metrics are used to calculate the best path for routing.

The IS-IS configuration for each router is prepared using the preceding template. It is covered in the next section.

Configuration

The configuration for each router is determined by applying the preceding template to the topology diagram.

The rendered configuration for each router is shown here:

- The following shows the IS-IS protocol configuration and how it is enabled for the loopback and physical interfaces on router PE1:

PE1

```
router isis IGP
is-type level-2-only
net 49.0000.0000.0001.00
log adjacency changes
address-family ipv4 unicast
  metric-style wide Level-2

interface Loopback0
  passive
  address-family ipv4 unicast

interface GigabitEthernet0/0/0/2
  circuit-type level-2-only
  point-to-point
  hello-padding disable
  address-family ipv4 unicast
   metric 10
```

- The IS-IS configuration for router P2 is as follows:

P2

```
router isis IGP
is-type level-2-only
net 49.0000.0000.0002.00
log adjacency changes
address-family ipv4 unicast
  metric-style wide Level-2

interface Loopback0
  passive
  address-family ipv4 unicast

interface GigabitEthernet0/0/0/0
  circuit-type level-2-only
  point-to-point
  hello-padding disable
  address-family ipv4 unicast
   metric 10

interface GigabitEthernet0/0/0/1
  circuit-type level-2-only
  point-to-point
  hello-padding disable
  address-family ipv4 unicast
   metric 10

interface GigabitEthernet0/0/0/2
  circuit-type level-2-only
  point-to-point
  hello-padding disable
  address-family ipv4 unicast
   metric 10

interface GigabitEthernet0/0/0/3
  circuit-type level-2-only
  point-to-point
  hello-padding disable
  address-family ipv4 unicast
   metric 10
```

- The IS-IS configuration for router P3 is as follows:

P3

```
router isis IGP
is-type level-2-only
net 49.0000.0000.0003.00
log adjacency changes
address-family ipv4 unicast
  metric-style wide Level-2

interface Loopback0
  passive
  address-family ipv4 unicast

interface GigabitEthernet0/0/0/0
  circuit-type level-2-only
  point-to-point
  hello-padding disable
  address-family ipv4 unicast
   metric 10

interface GigabitEthernet0/0/0/1
  circuit-type level-2-only
  point-to-point
  hello-padding disable
  address-family ipv4 unicast
   metric 10

interface GigabitEthernet0/0/0/2
  circuit-type level-2-only
  point-to-point
  hello-padding disable
  address-family ipv4 unicast
   metric 10
```

- The IS-IS configuration for router P4 is as follows:

P4

```
router isis IGP
is-type level-2-only
net 49.0000.0000.0004.00
log adjacency changes
address-family ipv4 unicast
  metric-style wide Level-2

interface Loopback0
```

```
  passive
  address-family ipv4 unicast

interface GigabitEthernet0/0/0/0
  circuit-type level-2-only
  point-to-point
  hello-padding disable
  address-family ipv4 unicast
   metric 10

interface GigabitEthernet0/0/0/1
  circuit-type level-2-only
  point-to-point
  hello-padding disable
  address-family ipv4 unicast
   metric 10

interface GigabitEthernet0/0/0/2
  circuit-type level-2-only
  point-to-point
  hello-padding disable
  address-family ipv4 unicast
   metric 10

interface GigabitEthernet0/0/0/4
  circuit-type level-2-only
  point-to-point
  hello-padding disable
  address-family ipv4 unicast
   metric 10
```

- The IS-IS configuration for router PE5 is as follows:

PE5

```
router isis IGP
is-type level-2-only
net 49.0000.0000.0005.00
log adjacency changes
address-family ipv4 unicast
  metric-style wide Level-2

interface Loopback0
  passive
  address-family ipv4 unicast

interface GigabitEthernet0/0/0/0
```

```
  circuit-type level-2-only
  point-to-point
  hello-padding disable
  address-family ipv4 unicast
   metric 10
```

- The IS-IS configuration for router P6 is as follows:

P6

```
router isis IGP
is-type level-2-only
net 49.0000.0000.0006.00
log adjacency changes
address-family ipv4 unicast
  metric-style wide Level-2

interface Loopback0
  passive
  address-family ipv4 unicast

interface GigabitEthernet0/0/0/0
  circuit-type level-2-only
  point-to-point
  hello-padding disable
  address-family ipv4 unicast
   metric 10

interface GigabitEthernet0/0/0/1
  circuit-type level-2-only
  point-to-point
  hello-padding disable
  address-family ipv4 unicast
   metric 10
```

- The IS-IS configuration for router P7 is as follows:

P7

```
router isis IGP
is-type level-2-only
net 49.0000.0000.0007.00
log adjacency changes
address-family ipv4 unicast
  metric-style wide Level-2

interface Loopback0
  passive
```

```
address-family ipv4 unicast

interface GigabitEthernet0/0/0/0
  circuit-type level-2-only
  point-to-point
  hello-padding disable
  address-family ipv4 unicast
   metric 10

interface GigabitEthernet0/0/0/1
  circuit-type level-2-only
  point-to-point
  hello-padding disable
  address-family ipv4 unicast
   metric 10

interface GigabitEthernet0/0/0/2
  circuit-type level-2-only
  point-to-point
  hello-padding disable
  address-family ipv4 unicast
   metric 10

interface GigabitEthernet0/0/0/3
  circuit-type level-2-only
  point-to-point
  hello-padding disable
  address-family ipv4 unicast
   metric 10

interface GigabitEthernet0/0/0/4
  circuit-type level-2-only
  point-to-point
  hello-padding disable
  address-family ipv4 unicast
   metric 10
```

- The IS-IS configuration for router P8 is as follows:

P8

```
router isis IGP
is-type level-2-only
net 49.0000.0000.0008.00
log adjacency changes
address-family ipv4 unicast
  metric-style wide Level-2
```

```
interface Loopback0
  passive
  address-family ipv4 unicast

interface GigabitEthernet0/0/0/1
  circuit-type level-2-only
  point-to-point
  hello-padding disable
  address-family ipv4 unicast
   metric 10

interface GigabitEthernet0/0/0/2
  circuit-type level-2-only
  point-to-point
  hello-padding disable
  address-family ipv4 unicast
   metric 10
```

After applying the preceding configuration, the routers should have end-to-end reachability, which is verified in the next section.

Verification

In IS-IS, there are three types of circuits: Level-1 (L1), Level-2 (L2), and Level-1-2 (L1-2). Each circuit type operates a specific area:

- **L1 Circuit**: Operates within a Level-1 area and carries information about networks within that area.

- **L2 Circuit**: Operates within a Level-2 area and carries information about networks in both L1 and L2 areas. The backbone area typically consists of routers interconnected in a contiguous L2 area.

- **L1-L2 Circuit**: Operates within both Level-1 and Level-2 areas and carries information about networks in both areas, as explained in the preceding points. The routers transmit separate Hello messages for each circuit type.

If two routers are in the Level-2 (L2) area, they can form an adjacency even if their Level-1 (L1) area IDs differ. This is because routers in a Level-2 area are only concerned with Level-2 information, and their adjacency formation is independent of Level-1 areas.

On the other hand, their Level-1 (L1) area IDs must match for adjacency formation in the Level-1 portion of the network.

In this topology, all routers are located within the Level-2 (L2) area.

With the preceding information, verify the adjacency status as done in the next section.

Verifying IS-IS adjacency

The following output from the PE1 router confirms that the IS-IS adjacency has been established and aligns with the configuration applied on the router. The same result is expected for the remaining links in the network:

```
RP/0/RP0/CPU0:PE1#show isis adjacency

IS-IS IGP Level-2 adjacencies:
System Id        Interface              SNPA            State Hold
Changed  NSF IPv4 IPv6

BFD   BFD
P2               Gi0/0/0/2              *PtoP*          Up     25    00:0
2:28 Yes None None

Total adjacency count: 1
RP/0/RP0/CPU0:PE1#
```

Here's how to understand the important details presented in the output:

- System Id: The system ID represents the unique identifier of the neighboring device with which the current device has formed an adjacency. In this case, the neighboring device has the system ID P2.

- Interface: This column shows the specific interface on the local device that is connected to the neighboring device with the specified system ID. In this case, the adjacency is formed on the Gi0/0/0/2 interface.

- SNPA: **Subnetwork Point of Attachment (SNPA)** is derived from the **Open Systems Interconnection (OSI)** model and corresponds to the data link layer (Layer 2) type of the neighboring device. In the output, the designation P2P indicates that the adjacency is established over a point-to-point connection between the local device and the neighboring device.

- State: This column shows the current state of the adjacency. In this case, the adjacency state is Up, which means the IS-IS adjacency has been successfully established, and the devices can exchange routing information.

- Hold Time: This column indicates the time remaining (in seconds) until the IS-IS adjacency is considered lost if no hello messages are received from the neighbor.

- Changed Time: The duration since the adjacency state was last modified, presented in the format of HH:MM:SS.

Once the adjacency comes up, the routers update link state information to each other, as verified in the next section.

Verifying IS-IS databases

IS-IS utilizes *flooding* as its mechanism, to ensure that the IS-IS LSDB of all the routers contains the IS-IS LSPDUs from every router present in the topology.

IS-IS uses two separate databases to store information about the network: one for Level-1 and another for Level-2 areas. However, in this book, the labs only make use of Level-2, as configured and shown here from the PE1 router:

```
RP/0/RP0/CPU0:PE1#show isis database

IS-IS IGP (Level-2) Link State Database
LSPID                 LSP Seq Num  LSP Checksum  LSP Holdtime/Rcvd  ATT/P/OL
PE1.00-00           * 0x00000006   0xac14        1045 /*             0/0/0
P2.00-00              0x00000008   0xdfd1        1056 /1200          0/0/0
P3.00-00              0x00000007   0x3ad5        1055 /1200          0/0/0
P4.00-00              0x00000009   0x458e        1056 /1200          0/0/0
PE5.00-00             0x00000006   0x92e2        1046 /1200          0/0/0
P6.00-00              0x00000007   0xda2d        1056 /1200          0/0/0
P7.00-00              0x00000007   0x856c        1056 /1200          0/0/0
P8.00-00              0x00000006   0xb918        1056 /1200          0/0/0

Total Level-2 LSP count: 8     Local Level-2 LSP count: 1
RP/0/RP0/CPU0:PE1#
```

Let's take a closer look at the link state information about router PE5 on PE1.

IS-IS uses **Type-Length-Value (TLV)** in its LSPDUs to convey essential information about the network's link state. Each TLV serves a specific purpose, such as indicating area ID, and more.

Understanding this LSPDU output will be extremely helpful for grasping the content in the upcoming chapters:

```
RP/0/RP0/CPU0:PE1#show isis database verbose internal PE5

IS-IS IGP (Level-2) Link State Database
LSPID                 LSP Seq Num  LSP Checksum  LSP Holdtime/Rcvd  ATT/P/
OL  LSP Length
PE5.00-00             0x00000006   0x92e2        1027 /1200          0/0/0   91
  TLV code:1 length:2
    Area Address:   49
  TLV code:22 length:17
    Metric: 10          IS-Extended P4.00
      SubTLV code:9 length:4
        Physical BW: 1000000 kbits/sec
  TLV code:129 length:1
    NLPID:          0xcc
  TLV code:132 length:4
```

```
     IP Address:        5.5.5.5
  TLV code:135 length:25
    Metric: 0              IP-Extended 5.5.5.5/32
      SubTLV code:4 length:1
        Prefix Attribute Flags: X:0 R:0 N:1 E:0 A:0
    Metric: 10             IP-Extended 45.0.0.0/24
      SubTLV code:4 length:1
        Prefix Attribute Flags: X:0 R:0 N:0 E:0 A:0
  TLV code:137 length:3
    Hostname:         PE5


Total Level-2 LSP count: 1     Local Level-2 LSP count: 0
RP/0/RP0/CPU0:PE1#
```

Let's break down the output:

- LSPID: This is the unique identifier for the LSPDU – in this case, PE5.00-00.

- LSP Seq Num: This is the sequence number of the LSPDU. It is used to track the latest version of the LSPDU. If a router receives an LSPDU with a higher sequence number than the one in its LSDB, the newer LSPDU is updated in the LSDB. If the sequence number is the same, the LSPDU is discarded, which helps ensure that only the latest information is retained in the LSDB.

- LSP Checksum: This is the checksum value for the LSPDU to ensure data integrity.

- LSP Holdtime/Rcvd: This is the remaining time (in seconds) before the LSPDU expires and is removed from the database and the LSPDU refresh interval.

- ATT/P/OL: These are the flags representing the LSPDU state: ATT = Attached, P = (not supported), OL = Overload (beyond the current scope).

- LSP Length: This is the length of the LSPDU in bytes.

- TLV (Type-Length-Value) codes: These are various TLVs that contain additional information within the LSPDU entry:

  - Area Addresses TLV (TLV Code: 1): This specifies the areas to which the router belongs. The number 49 is considered a private area, similar to private IP addresses.

  - Protocols Supported TLV (TLV Code: 129): Also, **NLPID (Network Layer Protocol Identifier)**, is used to identify the type of network layer protocol being carried in the payload. The 0xcc value in the NLPID field signifies that the payload contains information related to IPv4 routing. This allows IS-IS routers to recognize and interpret the encapsulated IPv4 routing data.

  - IP Interface Address TLV (TLV Code: 132): Indicates the IP address of the device, 5.5.5.5.

- Extended IP reachability TLV (TLV Code: 135): This TLV is used to carry IP reachability information in IS-IS LSPDUs. It allows IS-IS routers to exchange information about IP prefixes and their subnet masks, effectively extending IS-IS's capabilities to include IP route propagation and resulting in the injection of the route into the **Routing Information Base** (**RIB**). It also provides information about the metrics (cost) associated with specific IP addresses. It shows two entries:

  - Metric: 0 IP-Extended 5.5.5.5/32 with a metric of 0 for the IP address 5.5.5.5/32 (loopback).

  - Metric: 10 IP-Extended 45.0.0.0/24 with a metric of 10 for the IP address subnet 45.0.0.0/24.

- Dynamic Hostname TLV (TLV Code: 137): This indicates the hostname of the device, PE5.

- Extended IS reachability (TLV Code: 22): This TLV is used to advertise the presence of IS neighbors that are reachable via a certain IP subnet and provides information about the IS-extended metrics (wide metrics). These metrics go beyond the narrow metric range of up to 63. It shows one entry:

  - Metric: 10 IS-Extended P4.00 with a metric of 10 for the IS-IS neighbor with system ID P4.00.

Each router in the topology learns the link state information for every other router through flooding and then populates its own link state database. The next section covers how the best path or best route is calculated using this information.

Verifying IS-IS routing

IS-IS on each router runs the SPF algorithm on the LSDB independently. The SPF algorithm calculates the shortest path to every reachable destination within the network based on the information in the LSDB.

The SPF algorithm works by considering the cost associated with each link in the LSDB. The cost represents the time or effort required to traverse that link. In IS-IS, the cost of a path is composed of the sum of metrics associated with each link along that path. The algorithm then explores all possible paths from the source to every destination and selects the path with the lowest cumulative cost as the optimum route.

Once the SPF algorithm has completed its calculations, IS-IS updates its routing table with the best routes to each destination. These routes are chosen based on the shortest path, ensuring that data packets take the most efficient path to reach their destinations. By running the SPF algorithm and updating the routing table as the network changes, IS-IS maintains an up-to-date and optimized routing system.

A constrained SPF considers more attributes beyond the metric. This will be briefly explored in the TI-LFA part later in the book. The following output shows the IS-IS route from PE1 to PE5, calculated by SPF after processing the LSDB:

```
RP/0/RP0/CPU0:PE1#show isis route detail 5.5.5.5/32

L2 5.5.5.5/32 [40/115] Label: None, medium priority
   Installed Jan 01 06:02:54.493 for 00:02:56
      via 12.0.0.2, GigabitEthernet0/0/0/2, P2, Weight: 0
      src PE5.00-00, 5.5.5.5
RP/0/RP0/CPU0:PE1#
```

Let's take a closer look at the routing parts of the preceding output:

- L2 5.5.5.5/32: A route to the IP address 5.5.5.5/32 within the IS-IS Level-2 network.

- [40/115]: The overall metric from source to destination is 40 and in the context of Cisco IOS XR, the administrative distance attributed to the IS-IS protocol is set to 115. Every routing protocol is assigned an administrative distance. When multiple routing protocols compete to insert the same subnet into the routing table or RIB, the administrative distances come into play. The route with the lowest administrative distance is prioritized and selected as the best route.

- 12.0.0.2, GigabitEthernet0/0/0/2, P2: This information denotes the next-hop IP address for forwarding, the egress interface, and the system ID of the neighboring IS-IS router.

- src PE5.00-00: The IP address 5.5.5.5/32 originates from the IS-IS system ID labeled PE5.00-00.

After calculating the best path to a destination, the route is then injected into the routing table or the routing information base, covered next.

Verifying the RIB

The RIB is a data structure that holds the complete routing information for a router. It includes all the routing entries that have been learned from various sources, such as static routes, dynamic routing protocols (such as OSPF, BGP, and EIGRP), and directly connected networks. The RIB contains a comprehensive list of routes, their associated prefixes, next-hop IP addresses, routing metrics, administrative distances, and other relevant information.

The following output confirms that all the loopback and point-to-point links of every router are learned and installed in the routing table. Every router in the network should have these same routes in their routing table.

Except for the directly connected and local subnets, all IP addresses are learned through the IS-IS Level-2 protocol, which has a fixed administrative distance of 115 and an associated end-to-end cost, making the IP addresses reachable via the respective interfaces. The following command output shows the entire routing table of the RIB of PE1:

```
RP/0/RP0/CPU0:PE1#show route

Codes: C - connected, S - static, R - RIP, B - BGP, (>) - Diversion path
       D - EIGRP, EX - EIGRP external, O - OSPF, IA - OSPF inter area
       N1 - OSPF NSSA external type 1, N2 - OSPF NSSA external type 2
       E1 - OSPF external type 1, E2 - OSPF external type 2, E - EGP
       i - ISIS, L1 - IS-IS level-1, L2 - IS-IS level-2
       ia - IS-IS inter area, su - IS-IS summary null, * - candidate default
       U - per-user static route, o - ODR, L - local, G  - DAGR, l - LISP
       A - access/subscriber, a - Application route
       M - mobile route, r - RPL, t - Traffic Engineering, (!) - FRR Backup
path

Gateway of last resort is not set

L    1.1.1.1/32 is directly connected, 00:03:48, Loopback0
i L2 2.2.2.2/32 [115/10] via 12.0.0.2, 00:03:30, GigabitEthernet0/0/0/2
i L2 3.3.3.3/32 [115/20] via 12.0.0.2, 00:03:22, GigabitEthernet0/0/0/2
i L2 4.4.4.4/32 [115/30] via 12.0.0.2, 00:03:22, GigabitEthernet0/0/0/2
i L2 5.5.5.5/32 [115/40] via 12.0.0.2, 00:03:17, GigabitEthernet0/0/0/2
i L2 6.6.6.6/32 [115/20] via 12.0.0.2, 00:03:22, GigabitEthernet0/0/0/2
i L2 7.7.7.7/32 [115/20] via 12.0.0.2, 00:03:22, GigabitEthernet0/0/0/2
i L2 8.8.8.8/32 [115/30] via 12.0.0.2, 00:03:22, GigabitEthernet0/0/0/2
C    12.0.0.0/24 is directly connected, 00:03:48, GigabitEthernet0/0/0/2
L    12.0.0.1/32 is directly connected, 00:03:48, GigabitEthernet0/0/0/2
i L2 23.0.0.0/24 [115/20] via 12.0.0.2, 00:03:30, GigabitEthernet0/0/0/2
i L2 26.0.0.0/24 [115/20] via 12.0.0.2, 00:03:30, GigabitEthernet0/0/0/2
i L2 27.0.0.0/24 [115/20] via 12.0.0.2, 00:03:30, GigabitEthernet0/0/0/2
i L2 29.0.0.0/24 [115/20] via 12.0.0.2, 00:03:30, GigabitEthernet0/0/0/2
i L2 34.0.0.0/24 [115/30] via 12.0.0.2, 00:03:22, GigabitEthernet0/0/0/2
i L2 37.0.0.0/24 [115/30] via 12.0.0.2, 00:03:22, GigabitEthernet0/0/0/2
i L2 39.0.0.0/24 [115/30] via 12.0.0.2, 00:03:22, GigabitEthernet0/0/0/2
i L2 45.0.0.0/24 [115/40] via 12.0.0.2, 00:03:22, GigabitEthernet0/0/0/2
i L2 47.0.0.0/24 [115/30] via 12.0.0.2, 00:03:22, GigabitEthernet0/0/0/2
i L2 48.0.0.0/24 [115/40] via 12.0.0.2, 00:03:22, GigabitEthernet0/0/0/2
i L2 67.0.0.0/24 [115/30] via 12.0.0.2, 00:03:22, GigabitEthernet0/0/0/2
i L2 78.0.0.0/24 [115/30] via 12.0.0.2, 00:03:22, GigabitEthernet0/0/0/2
L    127.0.0.0/8 [0/0] via 0.0.0.0, 00:03:49
RP/0/RP0/CPU0:PE1#
```

The output confirms that IS-IS has calculated and injected routes to each and every other router in the network topology.

The preceding routing table output represents the RIB or control plane. Routers then download this information into the **Forwarding Information Base** (**FIB**) or data plane, which is responsible for forwarding traffic between routers.

Verifying the FIB

Cisco IOS XR software utilizes **Cisco Express Forwarding** (**CEF**) as the FIB or data plane mechanism to achieve optimal IP forwarding.

As seen here, the FIB has derived the reachability information for PE5 from the RIB, such as the destination prefix, outgoing interface, and the next-hop IP address:

RIB

```
RP/0/RP0/CPU0:PE1#show route 5.5.5.5/32

Routing entry for 5.5.5.5/32
  Known via "isis IGP", distance 115, metric 40, type level-2
  Installed Jan  1 06:02:54.493 for 00:03:39
  Routing Descriptor Blocks
    12.0.0.2, from 5.5.5.5, via GigabitEthernet0/0/0/2
      Route metric is 40
  No advertising protos.
RP/0/RP0/CPU0:PE1#
```

The route learned via IS-IS is injected into the routing table or RIB. This is then used to program the FIB, as shown next:

FIB

```
RP/0/RP0/CPU0:PE1#show cef 5.5.5.5/32 brief
5.5.5.5/32, version 147, internal 0x1000001 0x30 (ptr 0xe7f0198) [1], 0x600
(0xdfe2a80), 0xa28 (0xebfe1e8)
Updated Jan  1 06:02:57.949
remote adjacency to GigabitEthernet0/0/0/2
Prefix Len 32, traffic index 0, precedence n/a, priority 3
  via 12.0.0.2/32, GigabitEthernet0/0/0/2, 4 dependencies, weight 0, class 0
[flags 0x0]
    path-idx 0 NHID 0x0 [0x10400330 0x0]
    next hop 12.0.0.2/32
RP/0/RP0/CPU0:PE1#
```

The RIB is populated and the FIB is programmed. Next, the routers should be able to reach one another.

Verifying end-to-end IP reachability

After IS-IS has achieved convergence across all routers and each router has populated its routing table, it's time to confirm end-to-end communication. This validation involves running a ping test between remote networks.

A successful ping confirms that IS-IS IGP routing and forwarding are functioning as expected. All subnets participate in IS-IS, creating an end-to-end IP reachability network. A ping test from any source IP to any destination IP should also be successful:

```
RP/0/RP0/CPU0:PE1#ping 5.5.5.5
Type escape sequence to abort.
Sending 5, 100-byte ICMP Echos to 5.5.5.5, timeout is 2 seconds:
!!!!!
Success rate is 100 percent (5/5), round-trip min/avg/max = 6/13/38 ms
RP/0/RP0/CPU0:PE1#
```

The topology is ready for IP forwarding of traffic among all the routers, but MPLS requires labeled forwarding of traffic. The traditional MPLS protocol LDP, used in this chapter, is explored in the next section to enable labeled forwarding.

MPLS LDP

LDP is one of the protocols used in MPLS networks to establish label-switched paths (LSPs) and enable forwarding of labeled packets.

Here's an overview of how MPLS LDP works:

- **Neighbor Discovery and Session Establishment**:

    - LDP routers use a multicast address, 224.0.0.2, for neighbor discovery. They send LDP Hello messages to this multicast address on UDP port 646. These Hellos help routers discover potential LDP neighbors on the same subnet.

    - Once a router receives a Hello message from a potential neighbor, it establishes a TCP session with that neighbor using TCP port 646. This session is crucial for reliable communication between LDP peers.

- **Label Assignment**:

    - In an MPLS network, routers assign a unique label to each IP prefix or **Forwarding Equivalence Class** (**FEC**) injected into the routing protocol such as **Open Shortest Path First** (**OSPF**) or **IS-IS**.

    - Each label represents a specific destination network. All packets destined for a particular IP subnet can be grouped into one FEC.

- **Label Distribution**:

  - After the session establishment, LDP routers exchange label mapping messages over the established TCP connection. These messages contain label bindings, associating labels with specific network prefixes.

  - Each router advertises the labels it can assign for the prefixes it knows about. These labels can be either locally assigned (for IP prefixes in the router's routing table) or received from other LDP routers.

- **Label Information Base (LIB)**:

  - The LDP LIB contains both locally generated label bindings and those received from other LDP routers during label distribution.

  - It maintains a mapping between the FEC (IP prefixes in the routing table) and the corresponding labels that routers in the MPLS network will use for forwarding.

- **Label Forwarding Information Base (LFIB)**:

  - This is built from the LIB and local routing information.

  - The primary role of the LFIB is to map incoming labels to the appropriate outgoing interfaces and corresponding labels. When an MPLS router receives a labeled packet, it checks the incoming label against the LFIB to determine the next hop and the label that should replace the incoming label.

- **Label Switching Operations**:

  - **Push**: When a packet enters the MPLS network from a non-MPLS domain on an LDP router, the router adds (or pushes) an MPLS label onto the packet and forwards it based on the label.

  - **Swap**: In the MPLS core network, routers switch labels. They exchange the label on the incoming packet with a new one using information from the LFIB and then send the packet onward with the new label. This method allows the packet to move through the MPLS network without requiring an IP lookup.

  - **Pop**: When a labeled packet reaches its destination or exits the MPLS domain, the last hop router in the path removes the top label from the label stack. The router then forwards the packet based on the IP header's destination IP address, using traditional IP routing. To reduce processing on the destination router, the router immediately preceding the destination removes the label and forwards the IP packet directly to it. This process is known as **Penultimate Hop Popping (PHP)**.

- **Label Withdrawal and Control**: LDP routers maintain a dynamic environment, so if a router learns that a particular FEC is no longer reachable, it will send a label withdrawal message to its neighbors to inform them about the change.

- **Label Switched Path**: MPLS LSPs are unidirectionally established to direct traffic from the ingress router to the egress router. To facilitate bidirectional communication, separate LSPs are established for both the forward and reverse directions.

The next section quickly runs through the required configuration.

Configuration

The book is committed to a deep dive into segment routing intricacies. It opts for a simplified LDP-related configuration, departing from the usual practices in production networks, to ensure clarity. This deliberate choice expedites the application and removal of configurations, streamlining the learning journey throughout the book.

Hence, the next section demonstrates the simplistic way to configure it, which is applied to all the routers.

All routers

Apply the following configuration on all the routers in the network to enable MPLS LDP networking in the topology:

- The following enables MPLS LDP globally on the router and configures logging for LDP neighbor-related events. It means that the router will generate log messages related to the establishment and maintenance of LDP sessions with its neighboring routers:

```
mpls ldp
 log
  neighbor
```

- Enable `mpls oam`, a set of tools and protocols used for monitoring and troubleshooting MPLS networks. It helps detect and diagnose issues related to MPLS LSPs, ensuring that the network is functioning correctly and efficiently. The specification, initially introduced in RFC 4379 under the heading *Detecting Multi-Protocol Label Switched (MPLS) Data Plane Failures*, has undergone iterative updates to improve its standards and functionalities. The latest enhancements and refinements are reflected in RFC 8029:

```
mpls oam
```

- The following configuration activates MPLS LDP on interfaces within the IS-IS IPv4 address family:

```
router IS-IS IGP
 address-family ipv4 unicast
  mpls ldp auto-config
```

The next step is to verify that LDP is configured and set up correctly in the topology.

Verification

The LDP configuration is verified through a step-by-step process in the next sections to ensure label forwarding is achieved in the network topology.

Verfiying the LDP interface

The following output confirms that MPLS and LDP label distribution are now enabled on the core interface(s) participating in the IS-IS:

```
RP/0/RP0/CPU0:PE1#show mpls interfaces detail
Interface GigabitEthernet0/0/0/2:
        LDP(A)
        MPLS enabled
RP/0/RP0/CPU0:PE1#
```

All routers should display a similar output for their respective core links in the topology.

They initiate discovery as shown next.

Verifying LDP discovery

The following output confirms that LDP has discovered a neighbor in the local segment and neighbor details are also presented. All routers in the network should discover their IS-IS neighbors as LDP neighbors, confirming that mpls ldp auto-config is working as expected:

```
RP/0/RP0/CPU0:PE1#show mpls ldp discovery

Local LDP Identifier: 1.1.1.1:0
Discovery Sources:
  Interfaces:
    GigabitEthernet0/0/0/2 : xmit/recv
      VRF: 'default' (0x60000000)
      LDP Id: 2.2.2.2:0, Transport address: 2.2.2.2
          Hold time: 15 sec (local:15 sec, peer:15 sec)
          Established: Jan  1 06:02:33.737 (00:06:03 ago)


RP/0/RP0/CPU0:PE1#
```

The output indicates that the router with the local LDP identifier 1.1.1.1 has discovered an LDP neighbor with the LDP identifier 2.2.2.2 on the GigabitEthernet0/0/0/2 interface.

Verifying the LDP neighbor

The following output confirms that the LDP session with the discovered neighbor is up and running, and the neighbor has advertised its information over this session:

```
RP/0/RP0/CPU0:PE1#show mpls ldp neighbor

Peer LDP Identifier: 2.2.2.2:0
  TCP connection: 2.2.2.2:30136 - 1.1.1.1:646
  Graceful Restart: No
  Session Holdtime: 180 sec
  State: Oper; Msgs sent/rcvd: 30/31; Downstream-Unsolicited
  Up time: 00:06:08
  LDP Discovery Sources:
    IPv4: (1)
      GigabitEthernet0/0/0/2
    IPv6: (0)
  Addresses bound to this peer:
    IPv4: (6)
      2.2.2.2         12.0.0.2        23.0.0.2        26.0.0.2
      27.0.0.2        29.0.0.2
    IPv6: (0)

RP/0/RP0/CPU0:PE1#
```

Here's how to understand the important details presented in the output:

- `Peer LDP Identifier`: The LDP neighbor with the IP address 2.2.2.2 and a label space ID of 0. LDP uses this identifier to establish and maintain the label distribution session with the peer.

- `TCP connection`: The local device (PE1) is using the IP address 1.1.1.1 and an unreserved random port to communicate with the neighbor (2.2.2.2) on port 646, which is reserved for LDP.

- `Session Holdtime`: The remaining time (in seconds) until the LDP session will be considered expired or terminated if no keepalive messages are received. In this case, the hold time for the session is 180 seconds.

- `Up time`: The session up time starts counting from when the session was established.

- `LDP Discovery Sources: IPv4: (1)`: The number in parentheses indicates the count of discovery sources for the LDP neighbor. In this case, there is one discovery source.

- `GigabitEthernet0/0/0/2`: The interface (GigabitEthernet0/0/0/2) through which the LDP neighbor (2.2.2.2) was discovered.

- Addresses bound to this peer: IPv4: (6): The number in parentheses indicates the count of IPv4 addresses bound to this LDP neighbor. In this case, there are six bound IP addresses. The six IPv4 addresses (2.2.2.2, 12.0.0.2, 23.0.0.2, 26.0.0.2, 27.0.0.2, and 29.0.0.2) are associated with this LDP neighbor. Ignore 29.0.0.2 for now, it is observed here because ansible playbook execution configures all the IP addresses in the beginning but it is not in use until the router P9 is integrated in *Chapter 14*.

As the name implies, LDP allocates and distributes labels for each route in the IGP routing table after establishing neighborship, populating the label information base.

Verifying the MPLS Label Information Base (LIB)

The **Label Information Base (LIB)** stores the mapping between the FEC – in this case, IP prefixes and the corresponding labels that are assigned to those prefixes. This mapping is typically created and maintained by the label distribution protocol, which in this case is LDP.

By registering as a client with the **Label Switching Database (LSD)** locally on the router, the MPLS LDP obtains the capability to allocate labels. This ensures that the labels acquired lie within the designated range set by the LSD.

The following displays information about the clients registered with the MPLS LSD:

```
RP/0/RP0/CPU0:PE1#show mpls lsd clients
ID Services                           Location
-- ----------------------------------  ------------
0  LSD(A)                              0/RP0/CPU0
1  Static(A)                           0/RP0/CPU0
2  L2VPN(A)                            0/RP0/CPU0
3  LDP(A)                              0/RP0/CPU0
4  PIM:pim(A)                          0/RP0/CPU0
5  PIM6:pim6(A)                        0/RP0/CPU0
6  Application-Controller:XTC(A)       0/RP0/CPU0
7  BFD(A)                              0/RP0/CPU0
RP/0/RP0/CPU0:PE1#
```

The IOS-XR software, by default, allocates dynamic labels ranging from 24000 to 1048575:

```
RP/0/RP0/CPU0:PE1#show mpls label range
Range for dynamic labels: Min/Max: 24000/1048575
RP/0/RP0/CPU0:PE1#
```

The following output shows the MPLS labels allocated by the router. These labels come from the LSD and are assigned by the client LDP.

LDP allocates the labels from the same dynamic label range. The label range 0-15 in Cisco IOS XR is reserved for specific purposes, though they fall outside the scope of the book:

```
RP/0/RP0/CPU0:PE1#show mpls label table
Table Label   Owner                              State  Rewrite
----- ------- ------------------------------     ------ -------
0     0       LSD(A)                             InUse  Yes
0     1       LSD(A)                             InUse  Yes
0     2       LSD(A)                             InUse  Yes
0     13      LSD(A)                             InUse  Yes
0     24000   LDP(A)                             InUse  Yes
0     24001   LDP(A)                             InUse  Yes
0     24002   LDP(A)                             InUse  Yes
0     24003   LDP(A)                             InUse  Yes
0     24004   LDP(A)                             InUse  Yes
0     24005   LDP(A)                             InUse  Yes
0     24006   LDP(A)                             InUse  Yes
0     24007   LDP(A)                             InUse  Yes
0     24008   LDP(A)                             InUse  Yes
0     24009   LDP(A)                             InUse  Yes
0     24010   LDP(A)                             InUse  Yes
0     24011   LDP(A)                             InUse  Yes
0     24012   LDP(A)                             InUse  Yes
0     24013   LDP(A)                             InUse  Yes
0     24014   LDP(A)                             InUse  Yes
0     24015   LDP(A)                             InUse  Yes
0     24016   LDP(A)                             InUse  Yes
0     24017   LDP(A)                             InUse  Yes
0     24018   LDP(A)                             InUse  Yes
RP/0/RP0/CPU0:PE1#
```

Inspect the LDP **Label Information Base (LIB)** on PE1 to find the MPLS label associated with the loopback IP address of the PE5 router:

```
RP/0/RP0/CPU0:PE1#show mpls ldp bindings 5.5.5.5/32
5.5.5.5/32, rev 42
        Local binding: label: 24013
        Remote bindings: (1 peers)
              Peer                Label
              ----------------    ---------
              2.2.2.2:0           24009
RP/0/RP0/CPU0:PE1#
```

The vital details presented in the output are outlined as follows:

- 5.5.5.5/32: This shows the FEC (IP address) for which MPLS LDP bindings are displayed.

- Local binding: This indicates the label assigned by the local router (PE1) for the prefix 5.5.5.5/32. This label will be distributed to the upstream LDP neighbors in the network.

- Remote bindings: This indicates the label allocated and distributed by this downstream LDP neighbor, along with its neighbor ID.

In the Cisco IOS XR domain, MPLS LDP promptly assigns labels upon the injection of a route into the routing table. This implies that each router within the network must possess labels linked to all core subnets and loopback addresses.

Verifying the MPLS LFIB

The LFIB is a data structure used by MPLS-enabled routers to make forwarding decisions for labeled packets. It stores the mapping between incoming labels and the appropriate outgoing labels and interfaces. When an MPLS router receives a labeled packet, it looks up the LFIB to find the entry associated with the incoming label and determines the next hop and the next label for the packet.

While the LIB serves as a control-plane function, the LFIB operates within the data plane. Its primary purpose is to facilitate the forwarding of labeled packets between the routers along the label-switched path. By determining packet forwarding decisions based on incoming labels, the LFIB eliminates the necessity for IP lookups in the routing process.

The following output confirms that the remote bindings from the earlier output are now installed as the outgoing label in the LFIB:

```
RP/0/RP0/CPU0:PE1#show mpls forwarding prefix 5.5.5.5/32
Local  Outgoing    Prefix              Outgoing     Next Hop          Bytes
Label  Label       or ID               Interface                      Switched
------ ----------- ------------------- ------------ ---------------- ----------
--
24013  24009       5.5.5.5/32          Gi0/0/0/2    12.0.0.2          500
RP/0/RP0/CPU0:PE1#
```

Labeled traffic with a label the same as the local label will be forwarded along the outgoing interface to the next-hop IP address, after swapping it with the outgoing label.

The LFIB corresponds to the labeled traffic, where an incoming label is swapped with an outgoing label or the label is popped before forwarding.

After enabling MPLS forwarding, the FIB also updates with the labels. This means that a label is added to incoming IP packets in the MPLS domain, and subsequent routers, referring to their LFIB, forward the labeled packet along the LSP.

Verifying the FIB

Cisco Express Forwarding (CEF) also maintains an MPLS **Label Forwarding Database** (LFD). When an IP packet enters the MPLS network, a label is added, and the labeled packet is sent along the route.

The CEF table lookup shows the label imposed for a remote IP address, matching what's in the LFIB. The ingress router adds this label to the packet header. By doing this, the router ensures the packet is label-switched within the MPLS network. Subsequent routers use the label to make forwarding decisions based on entries in their LFIB.

For a specific IP address, all routers in the network have consistent labels installed in both the LFIB and FIB:

```
RP/0/RP0/CPU0:PE1#show cef 5.5.5.5/32 brief
5.5.5.5/32, version 147, internal 0x1000001 0x30 (ptr 0xe7f0198) [1], 0x600
(0xdfe2a80), 0xa28 (0xebfe1e8)
Updated Jan  1 06:02:57.949
remote adjacency to GigabitEthernet0/0/0/2
Prefix Len 32, traffic index 0, precedence n/a, priority 3
   via 12.0.0.2/32, GigabitEthernet0/0/0/2, 4 dependencies, weight 0, class 0
[flags 0x0]
   path-idx 0 NHID 0x0 [0x10400330 0x0]
   next hop 12.0.0.2/32
   remote adjacency
    local label 24013        labels imposed {24009}
RP/0/RP0/CPU0:PE1#
```

The next step is to verify the end-to-end MPLS reachability over the label-switched path.

Verifying end-to-end MPLS reachability

The mpls oam configuration was applied to activate tools essential for the detection and diagnosis of failures within the MPLS data plane, such as the ping and traceroute of label-switched paths along the network. This entails a comprehensive examination of label-switched paths throughout the network, offering troubleshooting capabilities using either IP addresses or labels for improved network visibility.

The following output confirms the establishment of an end-to-end LSP from PE1 to PE5. It provides details of all the intermediate hops in the path, along with the corresponding IP-to-label bindings utilized at each step:

```
RP/0/RP0/CPU0:PE1#traceroute mpls multipath ipv4 5.5.5.5/32 verbose source
1.1.1.1

Starting LSP Path Discovery for 5.5.5.5/32

Codes: '!' - success, 'Q' - request not sent, '.' - timeout,
  'L' - labeled output interface, 'B' - unlabeled output interface,
```

```
   'D' - DS Map mismatch, 'F' - no FEC mapping, 'f' - FEC mismatch,
   'M' - malformed request, 'm' - unsupported tlvs, 'N' - no rx label,
   'P' - no rx intf label prot, 'p' - premature termination of LSP,
   'R' - transit router, 'I' - unknown upstream index,
   'X' - unknown return code, 'x' - return code 0

Type escape sequence to abort.

LLL!
Path 0 found,
output interface GigabitEthernet0/0/0/2 nexthop 12.0.0.2
source 1.1.1.1 destination 127.0.0.1
  0 12.0.0.1 12.0.0.2 MRU 1500 [Labels: 24009 Exp: 0] multipaths 0
L 1 12.0.0.2 23.0.0.3 MRU 1500 [Labels: 24010 Exp: 0] ret code 8 multipaths 2
L 2 23.0.0.3 34.0.0.4 MRU 1500 [Labels: 24001 Exp: 0] ret code 8 multipaths 1
L 3 34.0.0.4 45.0.0.5 MRU 1500 [Labels: implicit-null Exp: 0] ret code 8
multipaths 1
! 4 45.0.0.5, ret code 3 multipaths 0
LL!
Path 1 found,
output interface GigabitEthernet0/0/0/2 nexthop 12.0.0.2
source 1.1.1.1 destination 127.0.0.0
  0 12.0.0.1 12.0.0.2 MRU 1500 [Labels: 24009 Exp: 0] multipaths 0
L 1 12.0.0.2 27.0.0.7 MRU 1500 [Labels: 24006 Exp: 0] ret code 8 multipaths 2
L 2 27.0.0.7 47.0.0.4 MRU 1500 [Labels: 24001 Exp: 0] ret code 8 multipaths 1
L 3 47.0.0.4 45.0.0.5 MRU 1500 [Labels: implicit-null Exp: 0] ret code 8
multipaths 1
! 4 45.0.0.5, ret code 3 multipaths 0

Paths (found/broken/unexplored) (2/0/0)
Echo Request (sent/fail) (7/0)
Echo Reply (received/timeout) (7/0)
Total Time Elapsed 101 ms
RP/0/RP0/CPU0:PE1#
```

It shows two discovered paths, labeled Path 0 and Path 1, with detailed information about each hop along the routes.

For each path, the output provides the IP addresses of the traversed nodes, the output interface used on the source router (e.g., GigabitEthernet0/0/0/2), and the associated MPLS labels. Additionally, it indicates the number of available hops for each path, offering potential load-balancing opportunities.

The return codes for each hop indicate the success or failure of the traceroute process, helping to identify possible issues, such as timeouts.

Examining the traceroute results

The preceding multipath traceroute output contains extensive information. Since it pertains to tracing an MPLS LSP, the focus will be on the labels and their operations throughout the packet's journey, as shown in *Figure 2.2*.

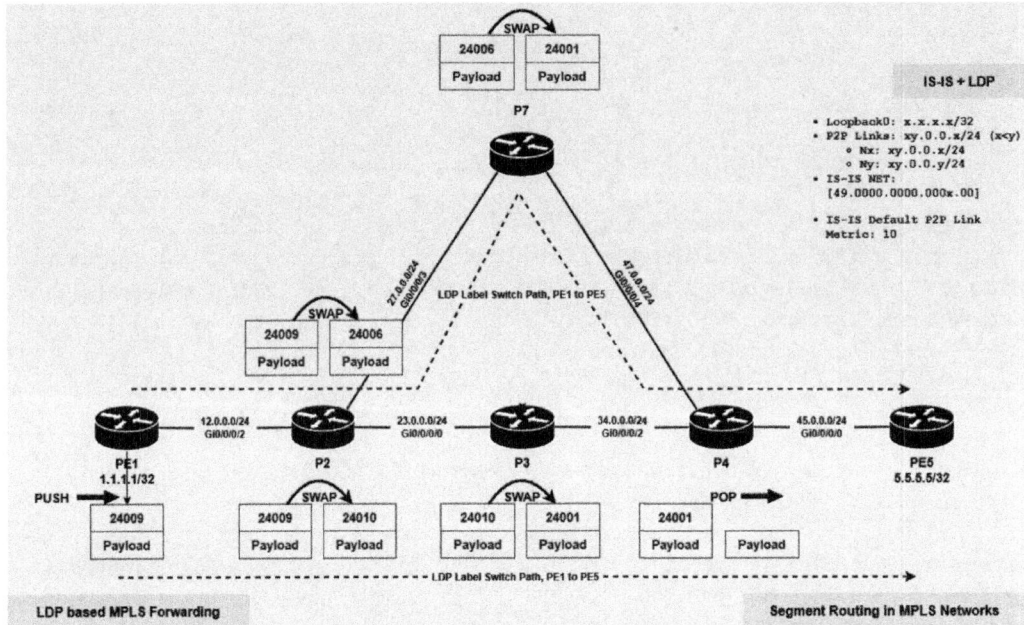

Figure 2.2 – LDP-based MPLS forwarding

Label PUSH

In an MPLS network, an IPv4 packet typically undergoes the following major operations, after entering and before exiting the MPLS domain:

- PE1 generates a local label for the destination router PE5 (5.5.5.5).

- P2, as one of the LDP neighbors, also distributes the local label it generated for the destination router PE5 to all its LDP neighbors, including PE1, immediately after injecting the route 5.5.5.5 into its IGP routing table.

- The route from PE1 to PE5 follows the IGP path through P2. The label received from P2, initially recorded in the LIB, is then programmed into both the LFIB and the FIB.

When an IPv4 packet enters the MPLS network at the ingress router (**Label Edge Router (LER)**), the router adds an MPLS label to the packet, creating an MPLS label stack. This label is inserted between the Ethernet header and the IPv4 header, as performed by PE1 acting as an LER. The MPLS header, including the outgoing label, is added before forwarding the packet into the MPLS domain – a process known as the PUSH operation – along the LSP.

> **Note**
>
> It's important to note that in scenarios such as MPLS-VPN or seamless-MPLS, there might be more than one label added, creating what's known as a label stack. The S-bit within the MPLS header serves the purpose of assisting the router in discerning whether it constitutes the final label within the stack or not.

```
RP/0/RP0/CPU0:PE1#show mpls forwarding prefix 5.5.5.5/32
Local  Outgoing    Prefix             Outgoing     Next Hop        Bytes
Label  Label       or ID              Interface                    Switched
------ ----------- ------------------ ------------ --------------- ----------
--
24013  24009       5.5.5.5/32         Gi0/0/0/2    12.0.0.2        500
RP/0/RP0/CPU0:PE1#
```

The preceding shows MPLS-to-MPLS LFIB programming and the following shows IP-to-MPLS forwarding programming:

```
RP/0/RP0/CPU0:PE1#show cef 5.5.5.5/32 brief
5.5.5.5/32, version 147, internal 0x1000001 0x30 (ptr 0xe7f0198) [1], 0x600
(0xdfe2a80), 0xa28 (0xebfe1e8)
Updated Jan  1 06:02:57.949
remote adjacency to GigabitEthernet0/0/0/2
Prefix Len 32, traffic index 0, precedence n/a, priority 3
   via 12.0.0.2/32, GigabitEthernet0/0/0/2, 4 dependencies, weight 0, class 0
[flags 0x0]
    path-idx 0 NHID 0x0 [0x10400330 0x0]
    next hop 12.0.0.2/32
    remote adjacency
     local label 24013      labels imposed {24009}
RP/0/RP0/CPU0:PE1#
```

Label SWAP

As the labeled packet traverses through the MPLS network, each MPLS router (LSR) examines the top label of the MPLS label stack and makes a forwarding decision based on the label's value. If an MPLS-labeled packet with an incoming label value matching a label in the local label table is encountered, it will be swapped with an outgoing label and forwarded to its neighbor. This process is referred to as the SWAP operation along the LSP, and the transit router responsible for this operation is also called an LSR.

Please note that LDP label forwarding in the network is influenced by the **Interior Gateway Protocol** (**IGP**) routing table. The following output displays two outgoing labels due to the **Equal Cost Multi-Path** (**ECMP**) feature, which is supported by default in IS-IS:

```
RP/0/RP0/CPU0:P2#show mpls forwarding prefix 5.5.5.5/32
Local   Outgoing    Prefix              Outgoing      Next Hop         Bytes
Label   Label       or ID               Interface                      Switched
------  ----------- ------------------- ------------  ---------------  ----------
--
24009   24010       5.5.5.5/32          Gi0/0/0/0     23.0.0.3         1628
        24006       5.5.5.5/32          Gi0/0/0/3     27.0.0.7         768
RP/0/RP0/CPU0:P2#
```

At each subsequent hop along the path, the label swap operation continues as shown here:

```
RP/0/RP0/CPU0:P3#show mpls forwarding prefix 5.5.5.5/32
Local   Outgoing    Prefix              Outgoing      Next Hop         Bytes
Label   Label       or ID               Interface                      Switched
------  ----------- ------------------- ------------  ---------------  ----------
--
24010   24001       5.5.5.5/32          Gi0/0/0/2     34.0.0.4         1252
RP/0/RP0/CPU0:P3#
```

Or, at any redundant next hop along the path:

```
RP/0/RP0/CPU0:P7#show mpls forwarding prefix 5.5.5.5/32
Local   Outgoing    Prefix              Outgoing      Next Hop         Bytes
Label   Label       or ID               Interface                      Switched
------  ----------- ------------------- ------------  ---------------  ----------
--
24006   24001       5.5.5.5/32          Gi0/0/0/4     47.0.0.4         512
RP/0/RP0/CPU0:P7#
```

Label POP

Finally, when the labeled packet reaches the penultimate hop before the egress router (LER), the MPLS label is removed (popped) from the packet using the POP operation, revealing the original IPv4 packet. Subsequently, the IPv4 packet is forwarded to its intended destination LER. This process is commonly referred to as **Penultimate Hop Popping** (**PHP**). This reduces the processing load on the egress router, allowing it to focus on IP packet processing.

When an MPLS router in Cisco IOS XR encounters an implicit NULL label (from the reserved range, label value 3), it indicates that this is the penultimate hop.

It is worth noting that for VPN traffic or packets with multiple labels stacked, only the topmost label is either swapped or popped, while the rest of the label stack remains intact as the packet is forwarded to its destination:

```
RP/0/RP0/CPU0:P4#show mpls forwarding prefix 5.5.5.5/32
Local  Outgoing    Prefix            Outgoing       Next Hop         Bytes
Label  Label       or ID             Interface                       Switched
------ ----------- ----------------- -------------- --------------- ----------
--
24001  Pop         5.5.5.5/32        Gi0/0/0/0      45.0.0.5         2584
RP/0/RP0/CPU0:P4#
```

The 5.5.5.5/32 IP address configured on PE5 prompts the generation of an implicit null label. This label is then advertised to LDP neighbor P4, signaling P4's role as the penultimate hop for PE5 and initiating a POP operation in the LFIB:

```
RP/0/RP0/CPU0:PE5#show mpls ldp bindings 5.5.5.5/32
5.5.5.5/32, rev 2
        Local binding: label: ImpNull
        Remote bindings: (1 peers)
            Peer                    Label
            ------------------      ---------
            4.4.4.4:0               24001
RP/0/RP0/CPU0:PE5#
```

This is the process of setting up and testing the MPLS LDP Label LSP for forwarding labeled traffic.

Tip

Replace the traceroute command with ping in the preceding command to test the MPLS ping.

Note

Since LSPs inherently operate in a unidirectional manner, the traceroute provided above illustrates the labeled path exclusively from PE1 to PE5.

It's important to note that for traffic moving from PE5 to PE1, a distinct LSP is established in the reverse direction. Perform an exercise to trace the LSP from PE5 to PE1 to better understand this.

To ensure the integrity of the end-to-end LSPs in the network, all routers should be able to perform both ping and traceroute operations with one another.

Summary

In this foundational chapter, the groundwork for the journey into segment routing with TI-LFA was established. Beginning with the construction of the base network topology, essential steps were taken to configure IP addresses on routers. The introduction of IS-IS as the interior gateway routing protocol enabled end-to-end IP routing. With the groundwork laid, the configuration of **Label Distribution Protocol (LDP)** was explored, facilitating end-to-end labeled forwarding within the MPLS network.

Throughout the lab, step-by-step configuration guidance was provided, emphasizing the importance of examining output results to validate configurations and providing an understanding of the underlying mechanisms of IS-IS and LDP protocols. It was stressed how important it is to check the results to make sure everything's working as it should.

While assuming a solid background in networking, the chapter ensured that even beginners could grasp the fundamentals by providing sufficient explanations of IS-IS and LDP protocols. This hands-on lab experience served as a crucial precursor to the exploration of segment routing with TI-LFA, empowering you to confidently navigate IS-IS and MPLS routing with assurance and expertise.

It's really important to practice this chapter well before moving on. The next chapters in the book focus on segment routing and build on the basics of IGP routing with IS-IS, label distribution, and labeled forwarding, which you learned about here.

References

Below are the key references cited throughout this chapter. These documents provide additional insights and technical details on MPLS, Fast Reroute, and Segment Routing, offering further reading for those interested in exploring these topics in depth.

- *[RFC0792] Postel, J., "Internet Control Message Protocol", STD 5, RFC 792, DOI 10.17487/ RFC0792, September 1981, <https://www.rfc-editor.org/rfc/rfc792>.*

- *[RFC1195] Callon, R., "Use of OSI IS-IS for routing in TCP/IP and dual environments", RFC 1195, DOI 10.17487/RFC1195, December 1990, <https://www.rfc-editor.org/info/rfc1195>.*

- *[RFC3031] Rosen, E., Viswanathan, A., and R. Callon, "Multiprotocol Label Switching Architecture", RFC 3031, DOI 10.17487/RFC3031, January 2001, <https://www.rfc-editor.org/info/rfc3031>.*

- *[RFC3032] Rosen, E., Tappan, D., Fedorkow, G., Rekhter, Y., Farinacci, D., Li, T., and A. Conta, "MPLS Label Stack Encoding", RFC 3032, DOI 10.17487/RFC3032, January 2001, <https://www.rfc-editor.org/info/rfc3032>.*

- *[RFC5036] Andersson, L., Ed., Minei, I., Ed., and B. Thomas, Ed., "LDP Specification", RFC 5036, DOI 10.17487/RFC5036, October 2007, <https://www.rfc-editor.org/info/rfc5036>.*

- *[RFC4950] Bonica, R., Gan, D., Tappan, D., and C. Pignataro, "ICMP Extensions for Multiprotocol Label Switching", RFC 4950, DOI 10.17487/RFC4950, August 2007, <https://www.rfc-editor.org/info/rfc4950>.*

- *[RFC5305] Li, T. and H. Smit, "IS-IS Extensions for Traffic Engineering", RFC 5305, DOI 10.17487/ RFC5305, October 2008, <https://www.rfc-editor.org/info/rfc5305>.*

- *[RFC5462] Andersson, L. and R. Asati, "Multiprotocol Label Switching (MPLS) Label Stack Entry: "EXP" Field Renamed to "Traffic Class" Field", RFC 5462, DOI 10.17487/RFC5462, February 2009, <https://www.rfc-editor.org/info/rfc5462>.*

- *[RFC7794] Ginsberg, L., Ed., Decraene, B., Previdi, S., Xu, X., and U. Chunduri, "IS-IS Prefix Attributes for Extended IPv4 and IPv6 Reachability", RFC 7794, DOI 10.17487/RFC7794, March 2016, <https://www.rfc-editor.org/info/rfc7794>.*

- *[RFC8029] Kompella, K., Swallow, G., Pignataro, C., Ed., Kumar, N., Aldrin, S., and M. Chen, "Detecting Multiprotocol Label Switched (MPLS) Data-Plane Failures", RFC 8029, DOI 10.17487/ RFC8029, March 2017, <https://www.rfc-editor.org/info/rfc8029>.*

- *[RFC5301] McPherson, D. and N. Shen, "Dynamic Hostname Exchange Mechanism for IS-IS", RFC 5301, DOI 10.17487/RFC5301, October 2008, <https://www.rfc-editor.org/info/rfc5301>.*

Part 2 - Segment Routing (SR-MPLS)

In this part, SR-MPLS is introduced into the topology, gradually shifting to an SR-MPLS-only configuration. The exploration extends to interactions between SR-MPLS-only and LDP-only domains within the MPLS network.

This part contains the following chapters:

- *Chapter 3, Lab 2 – Introducing Segment Routing MPLS (SR-MPLS)*
- *Chapter 4, Lab 3 – SR-LDP Interworking*

3

Lab 2 – Introducing Segment Routing MPLS (SR-MPLS)

The LDP-based MPLS network was fully operational by the end of the previous chapter, and we are now ready to introduce **Segment Routing MPLS** (**SR-MPLS**) into the topology. SR-MPLS will be introduced gradually in this chapter and without any operational impact on live traffic, as the same approach is expected to be used in the production network.

Objective

The objective of this lab is to introduce SR-MPLS into a traditional LDP-based MPLS topology. In the previous lab, emphasis was placed on examining the **Intermediate System-Intermediate System** (**IS-IS**) LSPDU and the LSDB, as understanding these elements is crucial for comprehending how **Segment Routing** (**SR**) information propagates throughout the network to establish the SR-MPLS control plane. While the MPLS data plane remains unchanged, the lab demonstrates how both MPLS control plane information types are programmed into two distinct MPLS data planes, a concept not dissimilar to the MPLS forwarding plane, as LDP and RSVP-TE also coexist in traditional MPLS networks.

This is achieved through the following tasks:

1. Design a straightforward schema for using **Segment Identifiers** (**SIDs**) in the SR-MPLS network, keeping it simple and easy to follow

2. Configure SR on a select few routers in the network

3. Inspect the IS-IS TLV extensions designed to facilitate the distribution of SR information

- Verify that SR-MPLS labels have been successfully installed in the MPLS forwarding plane

- Verify the end-to-end SR reachability within the SR-MPLS domain

- Verify that both LDP and SR-MPLS coexist in the network

- Verify the default forwarding preference of the **Forwarding Information Base** (**FIB**)

Before starting the lab, it is important to cover some theoretical details on SR-MPLS, which are presented in the next section.

Segment Routing MPLS (SR-MPLS)

The introduction of SR in an MPLS network does not require any changes to the underlying MPLS data plane or hardware forwarding capabilities.

In MPLS networks, routers use label-switching to forward packets based on pre-established **Label Switched Paths (LSPs)**. These labels are assigned by a control plane protocol such as **Label Distribution Protocol (LDP)** or **Resource Reservation Protocol with Traffic Engineering (RSVP-TE)**.

When SR is introduced, the source node determines and sets the entire path of a packet, from its origin to the final destination, including all the intermediate hops, using a segment or list of segments. But it doesn't replace the label-switching mechanism; instead, it uses the existing MPLS data plane and forwarding infrastructure to implement segment-based forwarding.

With SR-MPLS, the difference lies in how the labels are assigned, distributed, and utilized. The SR information is encoded as MPLS labels, and each label represents a specific segment or instruction that the packet should follow along its path. These segments can be represented by Prefix-SIDs, **Adjacency-SIDs (Adj-SID)**, or other special-purpose SIDs.

When a router receives a packet with an SR-MPLS label stack, it looks at the top label (topmost segment) to determine the next hop or operation to perform. It then forwards the packet based on the top label just like it would in a traditional MPLS network.

The advantage of SR-MPLS is that it provides greater flexibility and control over the packet forwarding path, as the segment information can be manipulated more easily by the network operators or through centralized control mechanisms. However, this flexibility is achieved without requiring any significant changes to the underlying MPLS forwarding hardware, making SR-MPLS deployment relatively straightforward and backward compatible with existing MPLS networks.

What is an SID?

The SIDs covered in this book are explained as follows:

In an SR-MPLS domain, the link-state IGP routing protocol propagates SR control information, which includes two types of segments, the **Prefix-SID** and the **Adjacency-SID**.

> **Note**
> This book doesn't cover other SIDs such as binding SIDs and anycast SIDs. It concentrates on the IGP segments.

The two types of segments propagated by IGP, also referred to as IGP segments, are explained here:

- **Prefix-SIDs as global segments**: Prefix-SIDs are a type of global segment used in SR-MPLS. Each Prefix-SID represents a specific network prefix and could be associated with a node (router) within the SR-MPLS domain. When a Prefix-SID is advertised, it indicates the presence of the associated prefix at the advertising router. This Prefix-SID allows the router to act as an ingress node for packets destined to the advertised prefix. The router uses the Prefix-SID encoding as a label to direct the traffic along the optimal path to reach the destination prefix.

  Prefix-SIDs are assigned from a global pool called the **Segment Routing Global Block (SRGB)**. Each router in the network configures and advertises these Prefix-SIDs as index values. Cisco IOS XR reserves a range of labels from 16000 to 23999, which provides 8,000 labels, specifically for SRGB. When SR-MPLS is activated on a link-state routing protocol, this label range becomes active, preventing clashes with dynamically assigned labels. Without this reservation, conflicts may arise, requiring a router reboot to address the issues.

  The Prefix-SID index is offset against 16000, and the resulting value is requested from the **Label Switch Database (LSD)** to be allocated as a label. All routers in the SR-MPLS domain receive this SRGB pool and index information through the IS-IS protocol, and they install the resulting label in their **Label Forwarding Information Base (LFIB)**. It is possible to configure a Prefix-SID with an absolute value, but that aspect won't be discussed in this book.

  When a Prefix-SID is associated with a loopback interface, it is also referred to as a node SID. Node SIDs must be globally unique within the SR-MPLS domain.

- **Adjacency-SIDs as local segments**: The Adjacency-SID represents a link or adjacency between two adjacent routers in the SR-MPLS network. It is assigned to the interface of a router that connects to its neighboring router. The purpose of the Adjacency-SID is to enable the router to forward packets directly to its adjacent neighbor.

  Unlike Prefix-SIDs, Adjacency-SIDs are dynamically allocated from the dynamic pool of the router and are advertised as absolute values in the IS-IS protocol. In Cisco IOS XR, the dynamic labels start from 24000.

  Only the router that allocates the Adjacency-SID installs it in its own LFIB. This makes the Adjacency-SID locally significant for that specific router.

Remote routers receive information about Adjacency-SIDs through the IS-IS link state packet, but they utilize it for traffic steering only when multiple labels are stacked together to route the traffic over a particular path.

In an MPLS LDP network, labels are primarily necessary to direct traffic from the ingress LER to the egress LER. Therefore, label allocation is specifically required for the loopback IP address of each node within the network. However, it's noteworthy that each router assigns and configures labels not only for the loopback addresses but also for every subnet within the network, encompassing all core links. It's important to note that while this is the typical behavior, some vendor implementations of LDP might not allocate labels for core links by default.

In SR, the Prefix-SIDs associated with loopback interfaces, often referred to as Node-SIDs, are programmed into the LFIB of all routers. This ensures proper label forwarding from the ingress LER to the egress LER. Meanwhile, the Adjacency-SID is solely installed into the locally originating router. This aligns with the fact that the topmost label of a labeled packet received on any router in the network will be either a Node-SID or its own Adjacency-SID. Before continuing with lab configuration, let's review the SR-MPLS configuration template.

Template

To begin with, introduce SR exclusively on the left half of the topology for the time being.

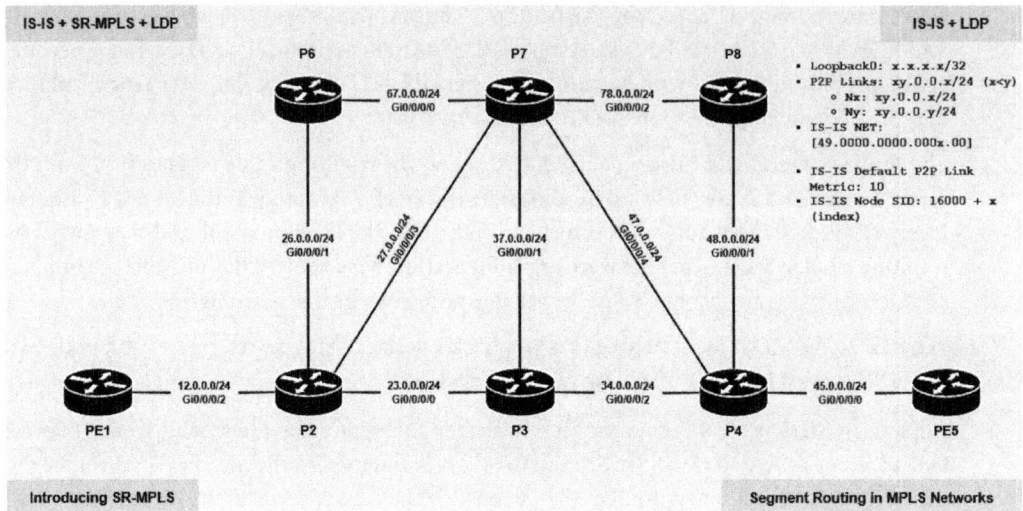

Figure 3.1 – Introducing SR-MPLS

As mentioned earlier, Cisco IOS XR is preconfigured with a default SRGB pool, spanning from 16000 to 23999. While it is possible to manually modify this range, the book adheres to the default setting.

Therefore, the sole configuration required involves setting the SID index value after enabling SR-MPLS, as follows:

```
router IS-IS IGP
address-family ipv4 unicast
  segment-routing mpls

interface Loopback0
  address-family ipv4 unicast
    prefix-sid index x
```

Let's break it down:

- `router IS-IS IGP`: This command enters the configuration section for the IS-IS routing protocol with the specified instance identifier or name: `IGP`.

- `address-family ipv4 unicast`: This command enters the IPv4 unicast address family for IS-IS.

- `segment-routing mpls`: This command enables SR using MPLS in the IS-IS configuration for IPv4 destinations.

- `interface Loopback0`: This command refers to the configuration of a logical loopback interface and enables IS-IS on it.

- `address-family ipv4 unicast`: This enters the IPv4 unicast address family context for the loopback interface.

- `prefix-sid index` **x**: This command assigns a Prefix-SID value to the loopback interface. `index` x indicates that the value of x is used to uniquely identify nodes in your network. Assign the value of x as the node ID, uniquely identifying each node. For instance, $x = 1$ represents PE1, $x = 2$ corresponds to P2, and so on.

Although not the primary focus of the book, it's important to recognize the manual setup of the SRGB through the following command:

```
segment-routing global-block 16000 23999
```

It is advisable to consistently maintain the SRGB throughout the network. This practice enhances clarity and facilitates ease of understanding and utilization.

Configuration

As stipulated in the template in the previous section, the following configuration enables SR-MPLS in IS-IS on the specified routers.

PE1

```
router IS-IS IGP
address-family ipv4 unicast
  segment-routing mpls

interface Loopback0
  address-family ipv4 unicast
    prefix-sid index 1
```

The IPv4 Prefix-SID index for PE1 is 1, for P2 is 2, and so forth.

P2

```
router IS-IS IGP
address-family ipv4 unicast
  segment-routing mpls

interface Loopback0
  address-family ipv4 unicast
   prefix-sid index 2
```

P3

```
router IS-IS IGP
address-family ipv4 unicast
  segment-routing mpls

interface Loopback0
  address-family ipv4 unicast
   prefix-sid index 3
```

P6

```
router IS-IS IGP
address-family ipv4 unicast
  segment-routing mpls

interface Loopback0
  address-family ipv4 unicast
   prefix-sid index 6
```

P7

```
router IS-IS IGP
address-family ipv4 unicast
  segment-routing mpls

interface Loopback0
  address-family ipv4 unicast
   prefix-sid index 7
```

The left side of the network in the topology diagram now has SR-MPLS configured along with the existing LDP from the previous chapter. The next verification steps will show how these two work together.

Verification

In the upcoming sections, the activation of SR on each router and the propagation of SR information throughout the network using the IS-IS IGP will be explored. This will result in the establishment of an SR-MPLS control plane and data plane network running parallel to the existing LDP network.

Verifying IS-IS LSD registration

As evident in the following snippet, activating SR-MPLS initiates IS-IS (IGP) registration as a client with LSD. This registration enables IS-IS to request label allocations from the router's available label space with ease:

```
RP/0/RP0/CPU0:PE1#show mpls lsd clients
ID Services                              Location
-- ------------------------------------- ------------
0  LSD(A)                                0/RP0/CPU0
1  Static(A)                             0/RP0/CPU0
2  L2VPN(A)                              0/RP0/CPU0
3  LDP(A)                                0/RP0/CPU0
4  PIM:pim(A)                            0/RP0/CPU0
5  PIM6:pim6(A)                          0/RP0/CPU0
6  Application-Controller:XTC(A)         0/RP0/CPU0
7  BFD(A)                                0/RP0/CPU0
8  ISIS:IGP(A)                           0/RP0/CPU0
RP/0/RP0/CPU0:PE1#
```

This is in addition to the other protocols already registered as LSD clients, such as LDP, which was discussed in the previous chapter.

Verifying SR-MPLS label allocation

After enabling SR-MPLS in IS-IS, the MPLS label table now shows the default SRGB range, reserving 8,000 labels for Prefix-SIDs, commencing at label 16000. Furthermore, all physical interfaces participating in IS-IS are assigned Adjacency-SIDs from the dynamic label range, represented as absolute values:

```
RP/0/RP0/CPU0:PE1#show mpls label table detail
Table Label   Owner                              State  Rewrite
----- ------- ---------------------------------- ------ -------
0     0       LSD(A)                             InUse  Yes
0     1       LSD(A)                             InUse  Yes
0     2       LSD(A)                             InUse  Yes
0     13      LSD(A)                             InUse  Yes
0     16000   ISIS(A):IGP                        InUse  No
   (Lbl-blk SRGB, vers:0, (start_label=16000, size=8000)
0     24000   LDP(A)                             InUse  Yes
   (IPv4, vers:0, 'default':4U, 2.2.2.2/32)
```

```
0      24001   LDP(A)                             InUse  Yes
       (IPv4, vers:0, 'default':4U, 29.0.0.0/24)
0      24002   LDP(A)                             InUse  Yes
       (IPv4, vers:0, 'default':4U, 27.0.0.0/24)
0      24003   LDP(A)                             InUse  Yes
       (IPv4, vers:0, 'default':4U, 26.0.0.0/24)
0      24004   LDP(A)                             InUse  Yes
       (IPv4, vers:0, 'default':4U, 23.0.0.0/24)
0      24005   LDP(A)                             InUse  Yes
       (IPv4, vers:0, 'default':4U, 3.3.3.3/32)
0      24006   LDP(A)                             InUse  Yes
       (IPv4, vers:0, 'default':4U, 39.0.0.0/24)
0      24007   LDP(A)                             InUse  Yes
       (IPv4, vers:0, 'default':4U, 37.0.0.0/24)
0      24008   LDP(A)                             InUse  Yes
       (IPv4, vers:0, 'default':4U, 34.0.0.0/24)
0      24009   LDP(A)                             InUse  Yes
       (IPv4, vers:0, 'default':4U, 4.4.4.4/32)
0      24010   LDP(A)                             InUse  Yes
       (IPv4, vers:0, 'default':4U, 47.0.0.0/24)
0      24011   LDP(A)                             InUse  Yes
       (IPv4, vers:0, 'default':4U, 45.0.0.0/24)
0      24012   LDP(A)                             InUse  Yes
       (IPv4, vers:0, 'default':4U, 48.0.0.0/24)
0      24013   LDP(A)                             InUse  Yes
       (IPv4, vers:0, 'default':4U, 5.5.5.5/32)
0      24014   LDP(A)                             InUse  Yes
       (IPv4, vers:0, 'default':4U, 6.6.6.6/32)
0      24015   LDP(A)                             InUse  Yes
       (IPv4, vers:0, 'default':4U, 67.0.0.0/24)
0      24016   LDP(A)                             InUse  Yes
       (IPv4, vers:0, 'default':4U, 7.7.7.7/32)
0      24017   LDP(A)                             InUse  Yes
       (IPv4, vers:0, 'default':4U, 78.0.0.0/24)
0      24018   LDP(A)                             InUse  Yes
       (IPv4, vers:0, 'default':4U, 8.8.8.8/32)
0      24020   ISIS(A):IGP                        InUse  Yes
       (SR Adj Segment IPv4, vers:0, index=1, type=0, intf=Gi0/0/0/2, nh=12.0.0.2)
0      24021   ISIS(A):IGP                        InUse  Yes
       (SR Adj Segment IPv4, vers:0, index=3, type=0, intf=Gi0/0/0/2, nh=12.0.0.2)
RP/0/RP0/CPU0:PE1#
```

Please note that in Cisco IOS XR, for each IS-IS adjacency, two Adjacency-SID labels are allocated, one with **Fast Reroute** (**FRR**) protection and the other without. However, the details about the difference between the two are beyond the scope of the book. The non-FRR Adjacency-SID will be used in backup paths in the upcoming chapters of the book:

```
RP/0/RP0/CPU0:PE1#show isis adjacency detail

IS-IS IGP Level-2 adjacencies:
System Id       Interface              SNPA          State Hold Changed  NSF
IPv4 IPv6
                                                                    BFD   BFD
P2              Gi0/0/0/2              *PtoP*        Up    28   00:19:32 Yes
None None
   Area Address:          49
   Neighbor IPv4 Address: 12.0.0.2*
   Adjacency SID:         24020
   Non-FRR Adjacency SID: 24021
   Topology:              IPv4 Unicast
   BFD Status:            BFD Not Required, Neighbor Useable

Total adjacency count: 1
RP/0/RP0/CPU0:PE1#
```

SR-MPLS is enabled and labels are allocated. The next step is to propagate this information throughout the network using the IS-IS IGP, as SR-MPLS relies on it.

Verifying the SR-MPLS IS-IS database

IS-IS TLVs have been extended based on RFC 8667, which is the *IS-IS Extensions for Segment Routing* specification. This expansion allows these **Type-Length-Value** (**TLV**) to carry SR information within the link-state PDUs.

The routers running SR expand their IS-IS **Link-State Protocol Data Units** (**LSPDUs**) by incorporating extra details via the extension of TLV structures with sub-TLVs. This extension enables the advertisement of SR information across the entire network.

To illustrate, the following output examines the LSPDU of PE1:

```
RP/0/RP0/CPU0:PE1#show isis database verbose internal PE1

IS-IS IGP (Level-2) Link State Database
LSPID                   LSP Seq Num  LSP Checksum  LSP Holdtime/Rcvd  ATT/P/
OL  LSP Length
PE1.00-00               * 0x00000009   0x1249         904  /*          0/0/0    154
   TLV code:1 length:2
     Area Address:    49
   TLV code:129 length:1
```

```
      NLPID:          0xcc
    TLV code:132 length:4
      IP Address:      1.1.1.1
    TLV code:135 length:33
      Metric: 10          IP-Extended 12.0.0.0/24
        SubTLV code:4 length:1
          Prefix Attribute Flags: X:0 R:0 N:0 E:0 A:0
      Metric: 0           IP-Extended 1.1.1.1/32
        SubTLV code:3 length:6
          Prefix-SID Index: 1, Algorithm:0, R:0 N:1 P:0 E:0 V:0 L:0
        SubTLV code:4 length:1
          Prefix Attribute Flags: X:0 R:0 N:1 E:0 A:0
    TLV code:137 length:3
      Hostname:         PE1
    TLV code:242 length:24
      Router Cap:     1.1.1.1 D:0 S:0
        SubTLV code:2 length:9
          Segment Routing: I:1 V:0, SRGB Base: 16000 Range: 8000
        SubTLV code:23 length:2
          Node Maximum SID Depth:
            Label Imposition: 10
        SubTLV code:19 length:2
          SR Algorithm:
            Algorithm: 0
            Algorithm: 1
    TLV code:22 length:46
      Metric: 10          IS-Extended P2.00
        SubTLV code:4 length:8
          Local Interface ID: 9, Remote Interface ID: 9
        SubTLV code:6 length:4
          Interface IP Address: 12.0.0.1
        SubTLV code:8 length:4
          Neighbor IP Address: 12.0.0.2
        SubTLV code:9 length:4
          Physical BW: 1000000 kbits/sec
        SubTLV code:31 length:5
          ADJ-SID: F:0 B:0 V:1 L:1 S:0 P:0 weight:0 Adjacency-sid:24021

 Total Level-2 LSP count: 1     Local Level-2 LSP count: 1
 RP/0/RP0/CPU0:PE1#
```

To understand the support for SR, let's dig into the relevant TLVs and closely examine the expanded sub-TLVs. The ensuing details are extracted from various IS-IS RFCs.:

- `TLV code:242`: **Router capability TLV** – The SR support capability is advertised by expanding the TLV as follows:

  - `S`: **Scope**. If the S bit is set (1), the IS-IS router capability TLV must be flooded across the entire routing domain. If the S bit is not set (0), the TLV *must not* be leaked between levels. This bit *must not* be altered during the TLV leaking.

  - `D`: **Down**. When the IS-IS router capability TLV is leaked from **Level 2 (L2)** to **Level 1 (L1)**, the D bit *must* be set. Otherwise, this bit *must* be clear. IS-IS router capability TLVs with the D bit set *must not* be leaked from L1 to L2. This is to prevent TLV looping.

  The scope and down bits are not set (0) in the output because the routers are operating in a single IS-IS L2 domain, and with no L1 routers, route leaking is not applicable.

  - `SubTLV code:2`; **SR capabilities sub-TLV** – The SR capabilities sub-TLV *must* be propagated throughout the level and *must not* be advertised across level boundaries:

    - `I`: **MPLS IPv4 flag**. If set, then the router is capable of processing SR-MPLS-encapsulated IPv4 packets on all interfaces.

    - `V`: **MPLS IPv6 flag**. If set, then the router is capable of processing SR-MPLS-encapsulated IPv6 packets on all interfaces.

    - **SRGB base**: `16000`; range: `8000`: The SID/label sub-TLV contains the first value of the SRGB while the range contains the number of SRGB elements. The range value *must* be higher than 0.

      IS-IS operates only for the IPv4 address family and not for IPv6, which is why the I bit is set (1) and the V bit is not set (0). The default SRGB range is also advertised, as IS-IS has successfully registered with the LSD.

  - `SubTLV code:19`; **SR algorithm sub-TLV** – The SR algorithm *must* be propagated throughout the level and *must not* be advertised across level boundaries. The SR algorithm sub-TLV allows the router to advertise the algorithms that the router is currently using.

    - **Algorithm 0**: SPF algorithm based on link metric. This is the well-known shortest-path algorithm as computed by the IS-IS decision process. Consistent with the deployed practice for link-state protocols, algorithm 0 permits any node to overwrite the SPF path with a different path based on the local policy.

    - **Algorithm 1**: Strict SPF algorithm based on link metric. The algorithm is identical to algorithm 0, but algorithm 1 requires that all nodes along the path honor the SPF routing decision. A local policy *must not* alter the forwarding decision computed by algorithm 1 at the node claiming to support algorithm 1.

      When the originating router does advertise the SR algorithm sub-TLV, then algorithm 0 *must* be present while non-zero algorithms *may* be present.

- `TLV code:135`; **extended IP reachability**: The TLV 135 is further extended as shown in the following to support Prefix-SID information:

  - `SubTLV code:3`; **Prefix-SID sub-TLV** – The Prefix-SID sub-TLV carries the SR IGP Prefix-SID as defined in RFC 8402. The Prefix-SID *must* be unique within a given IGP domain (when the L flag is not set).

    - R: **Re-advertisement flag**. If set, then the prefix to which this Prefix-SID is attached has been propagated by the router from either another level (i.e., from L1 to L2 or the opposite) or redistribution (e.g., from another protocol).

    - N: **Node SID flag**. If set, then the Prefix-SID refers to the router identified by the prefix. Typically, the N flag is set on Prefix-SIDs that are attached to a router loopback address. The N flag is set when the Prefix-SID is a Node-SID as described in RFC 8402.

    - P: **No Penultimate Hop-Popping (No-PHP) flag**. If set, then the penultimate hop *must not* pop the Prefix-SID before delivering the packet to the node that advertised the Prefix-SID.

    - E: **Explicit NULL flag**. If set, any upstream neighbor of the Prefix-SID originator *must* replace the Prefix-SID with a Prefix-SID that has an explicit NULL value (0 for IPv4 and 2 for IPv6) before forwarding the packet.

    - V: **Value flag**. If set, then the Prefix-SID carries a value (instead of an index). By default, the flag is unset.

    - L: **Local flag**. If set, then the value/index carried by the Prefix-SID has local significance. By default, the flag is unset.

    - **Algorithm**: Algorithm 0 allows any node to replace the SPF path with an alternative path determined by local policy. Any further information is beyond the scope of this book.

    - **Prefix-SID index**: The V flag and L flag are set to 0. The SID/Index/Label field is a four-octet index defining the offset in the SID/Label space advertised by this router.

- `TLV code:22`; **extended IS reachability**: The TLV 22 is further extended as shown in the following to support Adjacency-SID information:

  - `SubTLV code:31`; **Adjacency-SID sub-TLV** – The Adjacency-SID sub-TLV is an optional sub-TLV carrying the SR IGP Adjacency-SID as defined in RFC 8402 with flags and fields that may be used, in future extensions of SR, for carrying other types of SIDs:

    - F: **Address family flag**. If unset, then the Adjacency-SID is used when forwarding IPv4-encapsulated traffic to the neighbor. If set, then the Adjacency-SID is used when forwarding IPv6-encapsulated traffic to the neighbor.

    - B: **Backup flag**. If set, the Adjacency-SID is eligible for protection (e.g., using **IP Fast Reroute (IPFRR)** or MPLS-FRR as described in RFC 8402.

    - V: **Value flag**. If set, then the Adjacency-SID carries a value. By default, the flag is set.

- L: **Local flag**. If set, then the value/index carried by the Adjacency-SID has local significance. By default, the flag is set.

- S: **Set flag**. When set, the S flag indicates that the Adjacency-SID refers to a set of adjacencies (and therefore *may* be assigned to other adjacencies as well).

- P: **Persistent flag**. When set, the P flag indicates that the Adjacency-SID is persistently allocated, that is, the Adjacency-SID value remains consistent across router restart and/or interface flap.

- `Weight`: This value represents the weight of the Adjacency-SID for the purpose of load balancing. The use of the weight is defined in RFC 8402.

- `Adjacency-sid`: The V flag and L flag are set to 1. The SID/Index/Label field is a three-octet local label where the 20 rightmost bits are used for encoding the label value.

As observed in the preceding lists of TLVs, the router conveys SR-MPLS control-plane information (SR-MPLS label distribution) in the following manner:

- The SRGB through the router capability TLV extension

- The Prefix-SID via the extended IP reachability extension

- The Adjacency-SID using the extended IS reachability TLV extension, along with their associated flags, respectively

This fundamentally represents label distribution, in contrast to the traditional MPLS approach.

> **Note**
> Since SR label (segment) information is propagated through the IS-IS LSPDUs, routers that don't have SR enabled also receive this information. However, they simply ignore it because SR is not enabled on them, and they cannot process the additional information.

With the preceding information, all routers in the SR-MPLS domain begin creating tables to enable end-to-end traffic across the network to be labeled with SR.

Verifying the Prefix-SID table

After learning the Prefix-SIDs of all routers from the IS-IS link-state database, the following output shows how the local router establishes a mapping between the Prefix-SID and IP addresses among them:

```
RP/0/RP0/CPU0:PE1#show isis segment-routing label table

IS-IS IGP IS Label Table
Label          Prefix                  Interface
----------     ----------------        ---------
16001          1.1.1.1/32              Loopback0
```

```
16002         2.2.2.2/32
16003         3.3.3.3/32
16006         6.6.6.6/32
16007         7.7.7.7/32
RP/0/RP0/CPU0:PE1#
```

The preceding content aligns with the **Label Information Base** (**LIB**) from the preceding chapter, mirroring the characteristics of traditional MPLS.

The routing and forwarding tables shown next will explain how the Prefix-SIDs in this table were formed.

Verifying the Routing Information Base (RIB)

The IS-IS route indicates the presence of labels, specifically a local label assigned as 16007. This arises from PE1 utilizing the default SRGB offset, and the P7 router advertising its Prefix-SID/Node-SID with a value of 7. As a result, the label designated for outgoing traffic through P2 remains constant at 16007, aligning seamlessly with P2's default SRGB offset:

```
RP/0/RP0/CPU0:PE1#show isis route detail 7.7.7.7/32

L2 7.7.7.7/32 [20/115] Label: 16007, medium priority
    Installed Jan  1 06:19:49.201 for 00:05:55
      via 12.0.0.2, GigabitEthernet0/0/0/2, Label: 16007, P2, SRGB Base: 16000,
Weight: 0
        src P7.00-00, 7.7.7.7, prefix-SID index 7, R:0 N:1 P:0 E:0 V:0 L:0, Alg:0
RP/0/RP0/CPU0:PE1#
```

The following output illustrates the transformation of the IGP route into an SR-labeled route. This transformation is accompanied by the corresponding local and remote label mapping:

```
RP/0/RP0/CPU0:PE1#show route 7.7.7.7/32 detail

Routing entry for 7.7.7.7/32
  Known via "isis IGP", distance 115, metric 20, labeled SR, type level-2
  Installed Jan  1 06:19:49.201 for 00:05:55
  Routing Descriptor Blocks
    12.0.0.2, from 7.7.7.7, via GigabitEthernet0/0/0/2
      Route metric is 20
      Label: 0x3e87 (16007)
      Tunnel ID: None
      Binding Label: None
      Extended communities count: 0
      Path id:1        Path ref count:0
      NHID:0x2(Ref:20)
  Route version is 0xa (10)
  Local Label: 0x3e87 (16007)
```

```
  IP Precedence: Not Set
  QoS Group ID: Not Set
  Flow-tag: Not Set
  Fwd-class: Not Set
  Route Priority: RIB_PRIORITY_NON_RECURSIVE_MEDIUM (7) SVD Type RIB_SVD_TYPE_
LOCAL
  Download Priority 1, Download Version 261
  No advertising protos.
RP/0/RP0/CPU0:PE1#
```

The relevant SR-MPLS components of the preceding output are explained here:

- `labeled SR`: This shows that SR labels are in place for the destination IP address to steer the traffic.

- `Label: 0x3e87 (16007)`: This label is akin to the label acquired in traditional MPLS routing. It is produced by combining the SRGB offset obtained from the IS-IS next-hop, P2, with the Prefix-SID/Node-SID index received from the source, `P7`. `16000 (P2) + 7 (P7) = 16007`.

- `Local Label: 0x3e87 (16007)`: Corresponding to the local label used in traditional MPLS, this label is formed by adding the local SRGB offset to the Prefix-SID index received from the remote edge router, `16000 (PE1) + 7 (P7) = 16007`.

The local and remote labels are then programmed into the LFIB and the FIB, as shown next.

Verifying the Label Forwarding Information Base (LFIB)

The following output indicates that the Prefix-SIDs of all SR routers are installed in the MPLS forwarding table. Also, the Adjacency-SIDs is only installed for the local router, not for all routers:

```
RP/0/RP0/CPU0:PE1#show mpls forwarding
Local   Outgoing    Prefix             Outgoing      Next Hop        Bytes
Label   Label       or ID              Interface                     Switched
------  ----------- ------------------ ------------  --------------- ----------
--
16002   Pop         SR Pfx (idx 2)     Gi0/0/0/2     12.0.0.2        0
16003   16003       SR Pfx (idx 3)     Gi0/0/0/2     12.0.0.2        0
16006   16006       SR Pfx (idx 6)     Gi0/0/0/2     12.0.0.2        0
16007   16007       SR Pfx (idx 7)     Gi0/0/0/2     12.0.0.2        0
24000   Pop         2.2.2.2/32         Gi0/0/0/2     12.0.0.2        744
24001   Pop         29.0.0.0/24        Gi0/0/0/2     12.0.0.2        0
24002   Pop         27.0.0.0/24        Gi0/0/0/2     12.0.0.2        0
24003   Pop         26.0.0.0/24        Gi0/0/0/2     12.0.0.2        0
24004   Pop         23.0.0.0/24        Gi0/0/0/2     12.0.0.2        432
24005   24001       3.3.3.3/32         Gi0/0/0/2     12.0.0.2        0
24006   24002       39.0.0.0/24        Gi0/0/0/2     12.0.0.2        0
```

```
24007   24003      37.0.0.0/24        Gi0/0/0/2      12.0.0.2        0
24008   24004      34.0.0.0/24        Gi0/0/0/2      12.0.0.2        0
24009   24005      4.4.4.4/32         Gi0/0/0/2      12.0.0.2        0
24010   24006      47.0.0.0/24        Gi0/0/0/2      12.0.0.2        0
24011   24007      45.0.0.0/24        Gi0/0/0/2      12.0.0.2        0
24012   24008      48.0.0.0/24        Gi0/0/0/2      12.0.0.2        0
24013   24009      5.5.5.5/32         Gi0/0/0/2      12.0.0.2        500
24014   24010      6.6.6.6/32         Gi0/0/0/2      12.0.0.2        0
24015   24011      67.0.0.0/24        Gi0/0/0/2      12.0.0.2        0
24016   24012      7.7.7.7/32         Gi0/0/0/2      12.0.0.2        0
24017   24013      78.0.0.0/24        Gi0/0/0/2      12.0.0.2        0
24018   24014      8.8.8.8/32         Gi0/0/0/2      12.0.0.2        0
24020   Pop        SR Adj (idx 1)     Gi0/0/0/2      12.0.0.2        0
24021   Pop        SR Adj (idx 3)     Gi0/0/0/2      12.0.0.2        0
RP/0/RP0/CPU0:PE1#
```

Installing the Adjacency-SID of the remote routers is not required, as they will never be the top outgoing label in the label stack. This will be explored in detail later in the book.

As shown in the preceding table, on router PE1, for destination P7, the detailed inspection of its SR-MPLS label is displayed in the following code block. The Local and Outgoing labels are calculated based on the explanation given under *Verifying the Routing Information Base (RIB)*. The entry is specified as SR Pfx (idx 7), indicating it's an SR Prefix-SID with an index of 7:

```
RP/0/RP0/CPU0:PE1#show mpls forwarding labels 16007
Local   Outgoing   Prefix             Outgoing       Next Hop        Bytes
Label   Label      or ID              Interface                      Switched
------  ---------- ------------------ ------------ --------------- ----------
--
16007   16007      SR Pfx (idx 7)     Gi0/0/0/2      12.0.0.2        0
RP/0/RP0/CPU0:PE1#
```

The LFIB is referred to when MPLS-to-MPLS traffic forwarding is required. The FIB is used to switch, IP-to-MPLS traffic. Now that we have introduced SR-MPLS and LDP, the next section examines the behavior of the FIB.

Verifying the Forwarding Information Base (FIB)

The following output confirms that even though there's a new SR-MPLS path available, the PE1 router still prefers to use the LDP path in the FIB:

```
RP/0/RP0/CPU0:PE1#show cef 7.7.7.7/32 brief
7.7.7.7/32, version 208, labeled SR, internal 0x1000001 0x8130 (ptr 0xe7efe50)
[1], 0x600 (0xdfe34e8), 0xa28 (0xebfe328)
Updated Jan  1 06:19:49.213
remote adjacency to GigabitEthernet0/0/0/2
```

```
Prefix Len 32, traffic index 0, precedence n/a, priority 3
Extensions: context-label:16007
   via 12.0.0.2/32, GigabitEthernet0/0/0/2, 14 dependencies, weight 0, class 0
[flags 0x0]
    path-idx 0 NHID 0x0 [0x10400330 0x0]
    next hop 12.0.0.2/32
    remote adjacency
    local label 24016        labels imposed {24012}
RP/0/RP0/CPU0:PE1#
```

The LFIB supports multiple label forwarding entries for a single destination, accommodating different local labels from various LDPs that map to distinct outgoing labels.

In contrast, within the RIB, only one protocol can inject a route for the same destination, which is subsequently programmed into the FIB. Consequently, the FIB aligns with a singular set of local and outgoing labels distributed by one specific LDP.

The coexistence of an LDP and SR-MPLS LSPs from PE1 to P7 is illustrated in *Figure 3.2* and confirmed by the subsequent traceroute outputs.

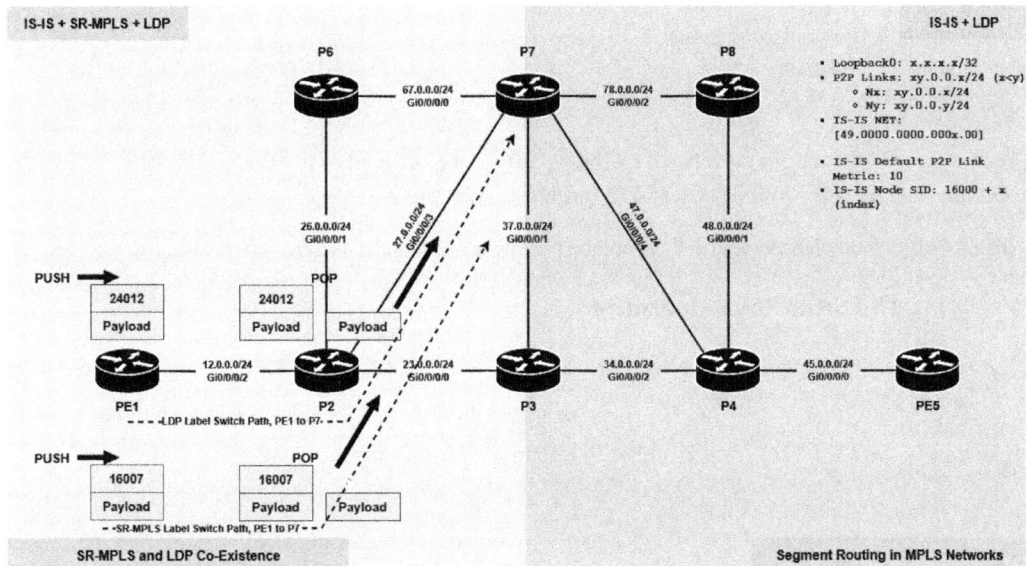

Figure 3.2 – SR-MPLS and LDP co-existence

By default, LDP takes precedence over SR-MPLS, resulting in the FIB continuing to reference LDP labels for IP-to-MPLS traffic switching.

Verifying the MPLS traceroute

The behavior of the FIB, as observed in the previous section, is confirmed by an MPLS traceroute from PE1 to P7:

```
RP/0/RP0/CPU0:PE1#traceroute mpls ipv4 7.7.7.7/32

Tracing MPLS Label Switched Path to 7.7.7.7/32, timeout is 2 seconds

Codes: '!' - success, 'Q' - request not sent, '.' - timeout,
   'L' - labeled output interface, 'B' - unlabeled output interface,
   'D' - DS Map mismatch, 'F' - no FEC mapping, 'f' - FEC mismatch,
   'M' - malformed request, 'm' - unsupported tlvs, 'N' - no rx label,
   'P' - no rx intf label prot, 'p' - premature termination of LSP,
   'R' - transit router, 'I' - unknown upstream index,
   'X' - unknown return code, 'x' - return code 0

Type escape sequence to abort.

  0 12.0.0.1 MRU 1500 [Labels: 24012 Exp: 0]
L 1 12.0.0.2 MRU 1500 [Labels: implicit-null Exp: 0] 15 ms
! 2 27.0.0.7 18 ms
RP/0/RP0/CPU0:PE1#
```

The usage of LDP labels is confirmed by the dynamic label instead of the Prefix-SID label 16007, as evidenced by its entry in the earlier LFIB output.

The SR-MPLS data plane validation can still be performed explicitly, as demonstrated in the next section.

Verifying the SR-MPLS traceroute

The SR-MPLS-specific ping and traceroute work successfully among all SR destinations, confirming end-to-end SR-MPLS reachability.

Like traditional MPLS LSP, it also allows **Penultimate Hop Popping** (**PHP**).

```
RP/0/RP0/CPU0:PE1#traceroute sr-mpls 7.7.7.7/32

Tracing MPLS Label Switched Path to 7.7.7.7/32, timeout is 2 seconds

Codes: '!' - success, 'Q' - request not sent, '.' - timeout,
   'L' - labeled output interface, 'B' - unlabeled output interface,
   'D' - DS Map mismatch, 'F' - no FEC mapping, 'f' - FEC mismatch,
   'M' - malformed request, 'm' - unsupported tlvs, 'N' - no rx label,
   'P' - no rx intf label prot, 'p' - premature termination of LSP,
   'R' - transit router, 'I' - unknown upstream index,
   'X' - unknown return code, 'x' - return code 0
```

```
Type escape sequence to abort.

  0 12.0.0.1 MRU 1500 [Labels: 16007 Exp: 0]
L 1 12.0.0.2 MRU 1500 [Labels: implicit-null Exp: 0] 20 ms
! 2 27.0.0.7 14 ms
RP/0/RP0/CPU0:PE1#
```

There are now two MPLS LSPs for the same IPv4 route in the LFIB: one utilizing LDP and the other using SR-MPLS.

This also implies that the **Label-Switching Routers** (**LSRs**) can route traffic for both the LSPs, whether they use LDP or SR-MPLS labels, based on the incoming labels.

Using SR-MPLS in FIB

In this chapter, the integration of SR into an existing traditional MPLS network has been explored. Two distinct methods are used to establish parallel label-switched paths. The FIB currently favors LDP labels over SR-MPLS labels, even though both LSPs are available. We will now adjust the FIB to prioritize SR for IP to MPLS traffic.

Configuration

The left half of the topology (*Figure 3.1*) is now capable of forwarding labeled traffic using both LDP and SR-MPLS for MPLS-to-MPLS traffic.

The default method for forwarding IP-to-MPLS traffic remains LDP, but it can be altered with the following configuration.

On SR-MPLS routers

The routers PE1, P2, P3, P6, and P7 within the SR-MPLS domain will now be configured with sr-prefer to prioritize SR-MPLS labels in the FIB:

```
router IS-IS IGP
address-family ipv4 unicast
  segment-routing mpls sr-prefer
```

sr-prefer is important only on the label edge routers. This is because the core routers will swap labels solely based on incoming labels. However, for completeness and the purpose of this book, it is applied to all routers that are configured to operate SR-MPLS.

Verification

The following steps will verify that the default method of forwarding labeled traffic has changed after adjusting the preference. By default, destinations using SR-MPLS will employ SR-MPLS labels, while those using LDP will utilize LDP labels.

Verifying FIB and MPLS traceroute to the SR and LDP destinations

The following output confirms that the FIB defaults to SR-MPLS labels for incoming IP packets entering the MPLS network:

```
RP/0/RP0/CPU0:PE1#show cef 7.7.7.7/32 brief
7.7.7.7/32, version 274, labeled SR, internal 0x1000001 0x8310 (ptr 0xe7efe50)
[1], 0x600 (0xdfe3530), 0xa28 (0xebfe3c8)
Updated Jan  1 06:29:13.034
remote adjacency to GigabitEthernet0/0/0/2
Prefix Len 32, traffic index 0, precedence n/a, priority 1
   via 12.0.0.2/32, GigabitEthernet0/0/0/2, 12 dependencies, weight 0, class 0
[flags 0x0]
    path-idx 0 NHID 0x0 [0x10400330 0x0]
    next hop 12.0.0.2/32
    remote adjacency
    local label 16007        labels imposed {16007}
RP/0/RP0/CPU0:PE1#
```

The MPLS traceroute follows suit accordingly:

```
RP/0/RP0/CPU0:PE1#traceroute mpls ipv4 7.7.7.7/32

Tracing MPLS Label Switched Path to 7.7.7.7/32, timeout is 2 seconds

Codes: '!' - success, 'Q' - request not sent, '.' - timeout,
  'L' - labeled output interface, 'B' - unlabeled output interface,
  'D' - DS Map mismatch, 'F' - no FEC mapping, 'f' - FEC mismatch,
  'M' - malformed request, 'm' - unsupported tlvs, 'N' - no rx label,
  'P' - no rx intf label prot, 'p' - premature termination of LSP,
  'R' - transit router, 'I' - unknown upstream index,
  'X' - unknown return code, 'x' - return code 0

Type escape sequence to abort.

  0 12.0.0.1 MRU 1500 [Labels: 16007 Exp: 0]
L 1 12.0.0.2 MRU 1500 [Labels: implicit-null Exp: 0] 18 ms
! 2 27.0.0.7 10 ms
RP/0/RP0/CPU0:PE1#
```

This step also indicates that the IP-to-MPLS traffic on the label edge router has transitioned from LDP to SR-MPLS. Basic ping tests on the Cisco IOS XRv9k platform show no packet loss during the migration from LDP to SR-MPLS.

Verifying MPLS traceroute to the LDP destinations

Meanwhile, the LDP continues to function as it always has:

```
RP/0/RP0/CPU0:PE1#traceroute mpls ipv4 5.5.5.5/32

Tracing MPLS Label Switched Path to 5.5.5.5/32, timeout is 2 seconds

Codes: '!' - success, 'Q' - request not sent, '.' - timeout,
  'L' - labeled output interface, 'B' - unlabeled output interface,
  'D' - DS Map mismatch, 'F' - no FEC mapping, 'f' - FEC mismatch,
  'M' - malformed request, 'm' - unsupported tlvs, 'N' - no rx label,
  'P' - no rx intf label prot, 'p' - premature termination of LSP,
  'R' - transit router, 'I' - unknown upstream index,
  'X' - unknown return code, 'x' - return code 0

Type escape sequence to abort.

  0 12.0.0.1 MRU 1500 [Labels: 24009 Exp: 0]
L 1 12.0.0.2 MRU 1500 [Labels: 24010 Exp: 0] 23 ms
L 2 23.0.0.3 MRU 1500 [Labels: 24001 Exp: 0] 13 ms
L 3 34.0.0.4 MRU 1500 [Labels: implicit-null Exp: 0] 17 ms
! 4 45.0.0.5 20 ms
RP/0/RP0/CPU0:PE1#
```

The router PE5, situated on the right side of the topology, does not support SR. Consequently, only the LDP-based LSP is available and utilized, as illustrated.

Summary

This chapter explored the integration of SR-MPLS into an existing MPLS network already using LDP. It outlined the process of enabling SR-MPLS using IS-IS as the IGP routing protocol. Examining the IS-IS LSDB reveals how SR information propagates across the network, providing insight into the control plane of SR-MPLS.

The IS-IS registration with the LSD, allocation of the default SRGB labels and dynamic Adjacency-SID labels, and propagation of SR information in IS-IS through TLVs breaks down the process of generating and distributing labels, highlighting the differences in how SR-MPLS operates compared to traditional MPLS.

A key concept introduced was SIDs, which are essentially like labels used in SR-MPLS for routing. Other key concepts, such as global and local segments (SIDs), were explained, with practical configuration and verification steps provided.

One significant aspect covered was the coexistence of SR-MPLS and traditional MPLS within the same network. The lab demonstrated how the SR-MPLS data plane operates alongside the traditional LDP data plane for labeled traffic. Verification processes involving RIB, FIB, and LFIB ensure both MPLS protocols function harmoniously, reaffirming default forwarding preferences.

Another noteworthy highlight of this chapter was that SR-MPLS was enabled alongside LDP. The next chapter explores the communication between SR-only and LDP-only domains.

References

Below are the key references cited throughout this chapter. These documents provide additional insights and technical details on MPLS, Fast Reroute, and Segment Routing, offering further reading for those interested in exploring these topics in depth.

- *[RFC3031] Rosen, E., Viswanathan, A., and R. Callon, "Multiprotocol Label Switching Architecture", RFC 3031, DOI 10.17487/RFC3031, January 2001, <https://www.rfc-editor.org/info/rfc3031>.*

- *[RFC3032] Rosen, E., Tappan, D., Fedorkow, G., Rekhter, Y., Farinacci, D., Li, T., and A. Conta, "MPLS Label Stack Encoding", RFC 3032, DOI 10.17487/RFC3032, January 2001, <https://www.rfc-editor.org/info/rfc3032>.*

- *[RFC5305] Li, T. and H. Smit, "IS-IS Extensions for Traffic Engineering", RFC 5305, DOI 10.17487/RFC5305, October 2008, <https://www.rfc-editor.org/info/rfc5305>.*

- *[RFC5307] Kompella, K., Ed. and Y. Rekhter, Ed., "IS-IS Extensions in Support of Generalized Multi-Protocol Label Switching (GMPLS)", RFC 5307, DOI 10.17487/RFC5307, October 2008, <https://www.rfc-editor.org/info/rfc5307>.*

- *[RFC5714] Shand, M. and S. Bryant, "IP Fast Reroute Framework", RFC 5714, DOI 10.17487/RFC5714, January 2010, <https://www.rfc-editor.org/info/rfc5714>.*

- *[RFC7981] Ginsberg, L., Previdi, S., and M. Chen, "IS-IS Extensions for Advertising Router Information", RFC 7981, DOI 10.17487/RFC7981, October 2016, <https://www.rfc-editor.org/info/rfc7981>.*

- *[RFC8029] Kompella, K., Swallow, G., Pignataro, C., Ed., Kumar, N., Aldrin, S., and M. Chen, "Detecting Multiprotocol Label Switched (MPLS) Data-Plane Failures", RFC 8029, DOI 10.17487/RFC8029, March 2017, <https://www.rfc-editor.org/info/rfc8029>.*

- *[RFC8402] Filsfils, C., Ed., Previdi, S., Ed., Ginsberg, L., Decraene, B., Litkowski, S., and R. Shakir, "Segment Routing Architecture", RFC 8402, DOI 10.17487/RFC8402, July 2018, <https://www.rfc-editor.org/info/rfc8402>.*

- [RFC8491] Tantsura, J., Chunduri, U., Aldrin, S., and L. Ginsberg, "Signaling Maximum SID Depth (MSD) Using IS-IS", RFC 8491, DOI 10.17487/RFC8491, November 2018, <https://www.rfc-editor.org/info/rfc8491>.

- [RFC8660] Bashandy, A., Ed., Filsfils, C., Ed., Previdi, S., Decraene, B., Litkowski, S., and R. Shakir, "Segment Routing with the MPLS Data Plane", RFC 8660, DOI 10.17487/RFC8660, December 2019, <https://www.rfc-editor.org/info/rfc8660>.

- [RFC8667] Previdi, S., Ed., Ginsberg, L., Ed., Filsfils, C., Bashandy, A., Gredler, H., and B. Decraene, "IS-IS Extensions for Segment Routing", RFC 8667, DOI 10.17487/RFC8667, December 2019, <https://www.rfc-editor.org/info/rfc8667>.

4

Lab 3 – SR-LDP Interworking

After the previous chapter, the network topology now employs two distinct MPLS forwarding mechanisms. The left side utilizes both SR-MPLS and LDP, while the right side operates solely on LDP. However, neither protocol utilizes the MPLS data-plane forwarding established by the other. Although the control-plane label distribution methods of SR-MPLS and LDP differ, their data-plane forwarding methods are the same. Both LDP and SR-MPLS eventually program the **Label Forwarding Information Base (LFIB)** and **Forwarding Information Base (FIB)** for MPLS-to-MPLS and IP-to-MPLS traffic forwarding, respectively. Hardware does not differentiate in forwarding traffic when utilizing one SR-MPLS label over the LDP label. This implies that routers running both SR-MPLS and LDP at the border of two domains should be capable of swapping incoming SR-MPLS labeled traffic with outgoing LDP labels, and vice versa.

Objective

In the previous chapter, only end-to-end forwarding utilizing either SR-MPLS LSP or LDP LSP was explored. The objective of this chapter is to initially divide the topology into two distinct MPLS domains: SR-MPLS only on the left-hand side and LDP only on the right-hand side of the diagram, as illustrated in *Figure 4.1*.

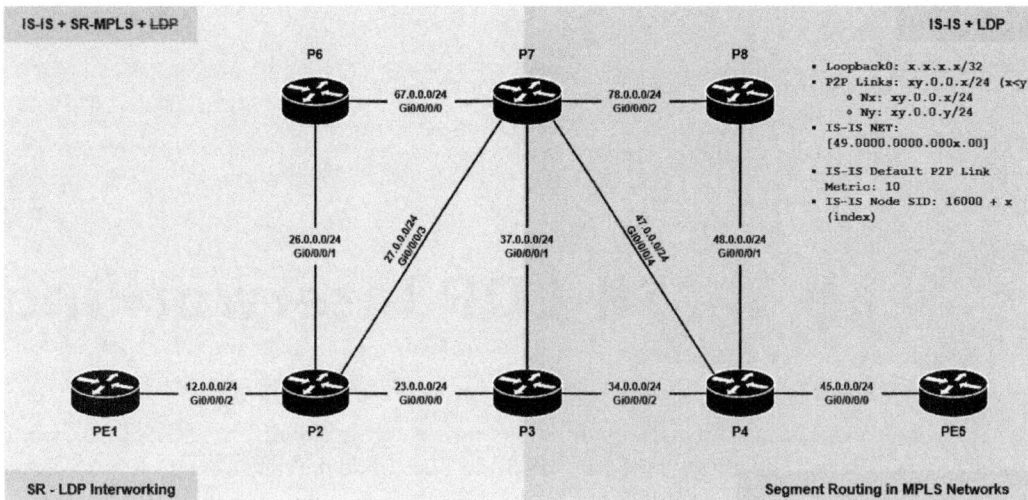

Figure 4.1 – SR-LDP interworking

The IS-IS IGP provides end-to-end routing among all the routers in the topology, making an end-to-end LSP desirable even if the routers are in two distinct MPLS domains with different label distribution methods.

This necessitates LSP stitching, where an incoming SR-MPLS label is mapped to an outgoing LDP label, and vice versa.

This is achieved through the tasks outlined here:

- Remove LDP on the left half of the network to make it SR-only

- Examine how **Label Switched Paths** (**LSPs**) stitch from LDP to SR

- Examine how LSPs stitch from SR to LDP using **Segment Routing Mapping Server** (**SRMS**)

- Confirm that label switch paths work seamlessly across different MPLS domains in both directions

- Complete the migration of the network topology from using LDP to SR-MPLS

At the end of this chapter, the entire topology is transformed into an SR-MPLS-only network.

MPLS domain separation

By removing LDP from routers PE1, P2, and P6, these routers are exclusively placed in the SR-MPLS domain. The routers P4, P8, and PE5 remain exclusively in the LDP domain. The routers P3 and P7 become the border routers, as they continue to run both SR-MPLS and LDP. This completes the separation of the network into two distinct MPLS domains: SR-MPLS and LDP.

The necessary configuration to achieve this will be shown now.

Configuration

The following configuration disables LDP label distribution on the routers.

PE1, P2, and P6

```
router isis IGP
address-family ipv4 unicast
  no mpls ldp auto-config
```

For ease of understanding and clarity, removing LDP from the IS-IS context is sufficient for the router to cease LDP functionality, as it will no longer distribute labels for IS-IS routing.

The LDP-based labeled forwarding would cease to function, as confirmed next.

Verification

The following steps confirm the absence of LDP labels and LDP-labeled paths on the local router to the LDP destination.

Verify MPLS LFIB

The routers PE1, P2, and P6 are no longer using LDP. PE5 is not part of the SR-MPLS domain, so from PE1, there is no end-to-end MPLS LSP to reach PE5.

Looking at the output on PE1, it is clear that the outgoing label entry is Unlabelled. This occurred because the LDP label assigned to PE5, received from P2, has been revoked as the LDP session is inactive/terminated following the removal of its configuration:

```
RP/0/RP0/CPU0:PE1#show mpls forwarding prefix 5.5.5.5/32
Local  Outgoing    Prefix              Outgoing      Next Hop        Bytes
Label  Label       or ID               Interface                     Switched
------ ----------- ------------------- ------------- --------------- ----------
--
24013  Unlabelled  5.5.5.5/32          Gi0/0/0/2     12.0.0.2        0
RP/0/RP0/CPU0:PE1#
```

Keep in mind that once an LDP local label is assigned by a router, it retains that assignment until the router is rebooted. This is the reason the local label is still in place.

Verify the MPLS traceroute

Due to the absence of an outgoing label on PE1 for PE5, the following output indicates that the traceroute request was not sent:

```
RP/0/RP0/CPU0:PE1#traceroute mpls ipv4 5.5.5.5/32

 Tracing MPLS Label Switched Path to 5.5.5.5/32, timeout is 2 seconds
```

```
Codes: '!' - success, 'Q' - request not sent, '.' - timeout,
  'L' - labeled output interface, 'B' - unlabeled output interface,
  'D' - DS Map mismatch, 'F' - no FEC mapping, 'f' - FEC mismatch,
  'M' - malformed request, 'm' - unsupported tlvs, 'N' - no rx label,
  'P' - no rx intf label prot, 'p' - premature termination of LSP,
  'R' - transit router, 'I' - unknown upstream index,
  'X' - unknown return code, 'x' - return code 0

Type escape sequence to abort.

  0 0.0.0.0 MRU 0 [No Label]
Q 1 *
RP/0/RP0/CPU0:PE1#
```

In simpler terms, it's established that there's no designated label switch path from PE1 to PE5.

The LSPs within each domain remain independent and do not interconnect with the other domain. To enable the exchange of a label distributed through LDP with an SR label, or vice versa, the LSPs from both domains need to be stitched together.

The unidirectional operation of LSPs necessitates the application of the stitching process separately in each direction.

The direction of LDP to SR stitching is explored in the next section.

LDP to SR stitching

The responsibility for stitching the labels between the two MPLS domains falls on the border routers that are running both protocols – namely, P3 and P7 in this case.

For the scenario examined currently, the source router is PE5 in the LDP domain and the destination router is PE1 in the SR-MPLS domain.

As soon as a route is injected into the RIB, the LDP allocates a local label for it and then distributes it to its downstream neighbors. Routers P3, P7, P4, P8, and PE5 perform this function even for routers PE1, P2, and P6, as all routers are within the same IS-IS routing domain.

This creates an LDP LSP from the router PE5 to P3 for the destination PE1.

At P3, upon realizing that no outgoing LDP label has been received from P2 for the destination PE1, the entry briefly remains unlabeled. However, P3 identifies that there is an SR-MPLS prefix-SID label from itself to PE1. Prefix then copies this SR-MPLS label over the previously unlabeled entry.

In the context of SR-LDP interworking, this process is known as the **Merge Operation**.

With this step, P3 successfully stitches the LDP LSP between PE5 and P3 with the SR-MPLS LSP between P3 and PE1, establishing an end-to-end MPLS LSP from source router PE5 to destination router PE1.

> **Note**
>
> On the Cisco IOS-XR platform, the merge operation automatically occurs swiftly as long as there is an outgoing label on the border router, without requiring any additional configuration or commands. Other platforms may require additional commands to enable this operation.

The LSP stitching and merge operation are verified and explained in the next section.

Verification

We will now verify the MPLS traceroute from PE5, which is running LDP, to PE1, which is running SR-MPLS.

Following that, we will confirm that LDP to SR stitching works seamlessly without requiring any additional configuration:

```
RP/0/RP0/CPU0:PE5#traceroute mpls multipath ipv4 1.1.1.1/32 verbose

Starting LSP Path Discovery for 1.1.1.1/32

Codes: '!' - success, 'Q' - request not sent, '.' - timeout,
   'L' - labeled output interface, 'B' - unlabeled output interface,
   'D' - DS Map mismatch, 'F' - no FEC mapping, 'f' - FEC mismatch,
   'M' - malformed request, 'm' - unsupported tlvs, 'N' - no rx label,
   'P' - no rx intf label prot, 'p' - premature termination of LSP,
   'R' - transit router, 'I' - unknown upstream index,
   'X' - unknown return code, 'x' - return code 0

Type escape sequence to abort.

LLL!
Path 0 found,
output interface GigabitEthernet0/0/0/0 nexthop 45.0.0.4
source 45.0.0.5 destination 127.0.0.0
   0 45.0.0.5 45.0.0.4 MRU 1500 [Labels: 24003 Exp: 0] multipaths 0
L 1 45.0.0.4 47.0.0.7 MRU 1500 [Labels: 24005 Exp: 0] ret code 8 multipaths 2
L 2 47.0.0.7 27.0.0.2 MRU 1500 [Labels: 16001 Exp: 0] ret code 8 multipaths 1
L 3 27.0.0.2 12.0.0.1 MRU 1500 [Labels: implicit-null Exp: 0] ret code 8
multipaths 1
! 4 12.0.0.1, ret code 3 multipaths 0
LL!
```

```
Path 1 found,
output interface GigabitEthernet0/0/0/0 nexthop 45.0.0.4
source 45.0.0.5 destination 127.0.0.1
  0 45.0.0.5 45.0.0.4 MRU 1500 [Labels: 24003 Exp: 0] multipaths 0
L 1 45.0.0.4 34.0.0.3 MRU 1500 [Labels: 24001 Exp: 0] ret code 8 multipaths 2
L 2 34.0.0.3 23.0.0.2 MRU 1500 [Labels: 16001 Exp: 0] ret code 8 multipaths 1
L 3 23.0.0.2 12.0.0.1 MRU 1500 [Labels: implicit-null Exp: 0] ret code 8
multipaths 1
! 4 12.0.0.1, ret code 3 multipaths 0

Paths (found/broken/unexplored) (2/0/0)
Echo Request (sent/fail) (7/0)
Echo Reply (received/timeout) (7/0)
Total Time Elapsed 149 ms
RP/0/RP0/CPU0:PE5#
```

Both paths in the traceroute output above demonstrate that the dynamic LDP label is swapped with the SR-MPLS Prefix-SID label on routers P7 and P3. Due to **Equal Cost Multiple-Path (ECMP)**, there are two available paths from PE5 to PE1, one via P7 and another via P3.

The diagram in *Figure 4.2* illustrates the traceroute operation described.

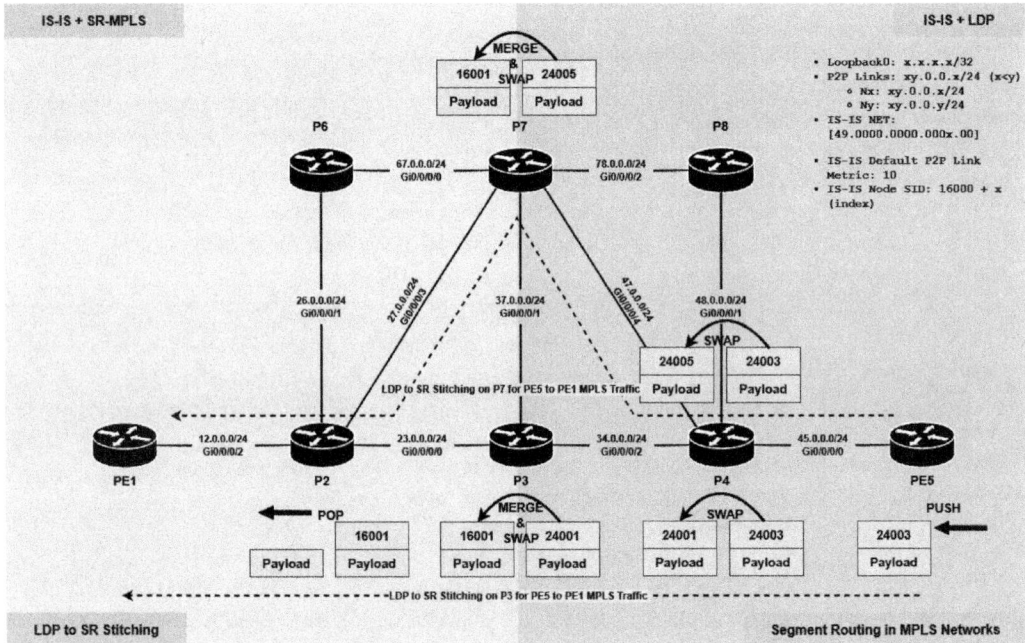

Figure 4.2 – LDP to SR stitching

This label-swapping process is examined in detail in the next section.

Examine the LFIB throughout the traceroute

The label-switching process is demonstrated in the following manner for the path going through P3 only, as shown in *Figure 4.3*:

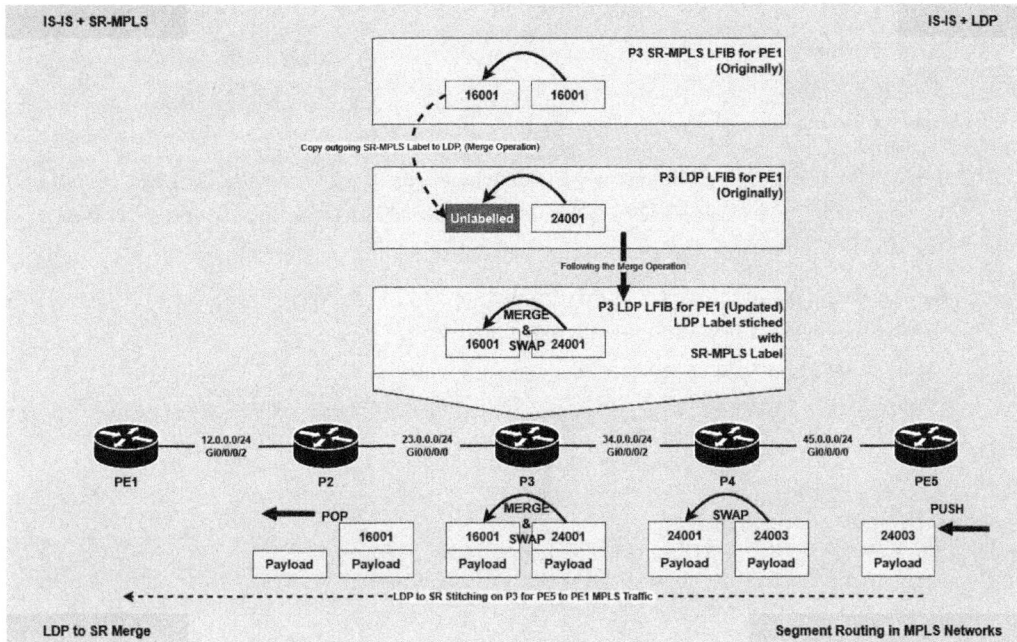

Figure 4.3 – LDP to SR merge operation

The routers P3, P4, and PE5, running LDP, consistently assign labels to all routes in their routing tables and distribute them to their LDP neighbors. As a result, there is an LDP path from PE5 to PE1, to P3, in this direction.

The hop-by-hop explanation is as follows:

- **Hop 0**: Outgoing label from PE5 to P4 using the LDP label.

    Initially, determine the MPLS label assigned to PE1 (1.1.1.1) on PE5. This label is obtained from the neighboring node (next-hop) P4 and serves as the outgoing label for that connection:

```
RP/0/RP0/CPU0:PE5#show mpls forwarding prefix 1.1.1.1/32
Local   Outgoing     Prefix              Outgoing     Next Hop         Bytes
Label   Label        or ID               Inter-
face                     Switched
------  -----------  ------------------  -----------  ---------------  -----
-------
24003   24003        1.1.1.1/32          Gi0/0/0/0    45.0.0.4         0
RP/0/RP0/CPU0:PE5#
```

The MPLS forwarding table can be checked using the known local label, as illustrated here:

```
RP/0/RP0/CPU0:PE5#show mpls forwarding labels 24003
Local  Outgoing    Prefix              Outgoing      Next Hop          Bytes
Label  Label       or ID               Inter-
face                            Switched
------ ----------- ------------------- ------------- ----------------- -----
-------
24003  24003       1.1.1.1/32          Gi0/0/0/0     45.0.0.4          0
RP/0/RP0/CPU0:PE5#
```

- **Hop 1**: The incoming LDP label on P4 is matched against the local label in the LFIB and then swapped with the outgoing LDP label to P3 (P3, which is a border router, operates both LDP and SR-MPLS).

   Keep querying the outgoing labels on the next hops from which they are learned. This will help identify the complete end-to-end label-switched path:

```
RP/0/RP0/CPU0:P4#show mpls forwarding labels 24003
Local  Outgoing    Prefix              Outgoing      Next Hop          Bytes
Label  Label       or ID               Inter-
face                            Switched
------ ----------- ------------------- ------------- ----------------- -----
-------
24003  24005       1.1.1.1/32          Gi0/0/0/4     47.0.0.7          1211
       24001       1.1.1.1/32          Gi0/0/0/2     34.0.0.3          384
RP/0/RP0/CPU0:P4#
```

   The following steps can be applied to P7 as well, given the existence of ECMPs from P4 to PE1.

- **Hop 2**: Label switched from P3 to P2 using SR-MPLS Label. (P2 doesn't run LDP, so it doesn't distribute any LDP label to P3. However, it operates SR-MPLS, enabling P3 to know P2's SRGB and PE1's Prefix-sid – used to reach PE1.)

   At P3, the displayed outgoing label represents the prefix-sid for PE1:

```
RP/0/RP0/CPU0:P3#show mpls forwarding labels 24001
Local  Outgoing    Prefix              Outgoing      Next Hop          Bytes
Label  Label       or ID               Inter-
face                            Switched
------ ----------- ------------------- ------------- ----------------- -----
-------
24001  16001       1.1.1.1/32          Gi0/0/0/0     23.0.0.2          256
RP/0/RP0/CPU0:P3#
```

This occurred because there isn't an outgoing LDP label from P3 to PE1; it's marked as Unla-belled. This is confirmed by the following output from the LDP forwarding table:

```
RP/0/RP0/CPU0:P3#show mpls ldp forwarding 1.1.1.1/32

Codes:
```

```
  - = GR label recovering, (!) = LFA FRR pure backup path
  {} = Label stack with multi-line output for a routing path
  G = GR, S = Stale, R = Remote LFA FRR backup
  E = Entropy label capability


Prefix          Label   Label(s)        Outgoing      Next
Hop             Flags
                In      Out             Inter-
face                            G S R E
--------------- ------- --------------- ------------ ------------------
-------
1.1.1.1/32      24001   Unlabelled      Gi0/0/0/0    23.0.0.2

RP/0/RP0/CPU0:P3#
```

Yet, P3 has an SR-MPLS Prefix-sid label assigned to PE1. This causes it to copy the outgoing Prefix-sid to the Unlabelled LFIB entry mentioned earlier. This automatic and immediate process is also called a **merge operation**, but it happens so quickly that you can't observe it in action:

```
RP/0/RP0/CPU0:P3#show mpls forwarding prefix 1.1.1.1/32
Local  Outgoing    Prefix              Outgoing      Next Hop        Bytes
Label  Label       or                                                Switched
ID                 Interface
------ ----------- ------------------- ------------ --------------- -----
-------
16001  16001       SR Pfx (idx 1)      Gi0/0/0/0    23.0.0.2        0
RP/0/RP0/CPU0:P3#


RP/0/RP0/CPU0:P3#show mpls forwarding labels 16001
Local  Outgoing    Prefix              Outgoing      Next Hop        Bytes
Label  Label       or                                                Switched
ID                 Interface
------ ----------- ------------------- ------------ --------------- -----
-------
16001  16001       SR Pfx (idx 1)      Gi0/0/0/0    23.0.0.2        0
RP/0/RP0/CPU0:P3#
```

• **Hop 3**: Label switched from P2 to PE1 using implicit-null PHP.

  P2 removes the label, and then the IP packet is sent toward its destination:

```
RP/0/RP0/CPU0:P2#show mpls forwarding labels 16001
Local  Outgoing    Prefix              Outgoing      Next Hop        Bytes
Label  Label       or                                                Switched
ID                 Interface
------ ----------- ------------------- ------------ --------------- -----
-------
16001  Pop         SR Pfx (idx 1)      Gi0/0/0/2    12.0.0.1        256
RP/0/RP0/CPU0:P2#
```

- **Hop 4**: Final destination at PE1.

The described label-switching process is the reason why the traceroute from the LDP router, PE5, to the SR router, PE1, succeeded smoothly without needing any extra steps.

The FIB is examined next to verify the same operation.

Verify the FIB

The merge operation is also observed in the FIB. The entry indicates that a local LDP label is swapped with an outgoing SR-MPLS label:

```
RP/0/RP0/CPU0:P3#show cef mpls local-label 24001 non-EOS brief
Label/EOS 24001/0, Label-type Unknown, version 235, labeled SR, internal
0x1000001 0x87f0 (ptr 0xe1be6e0) [1], 0x600 (0xdfe42b0), 0xa28 (0xebfe288)
Updated Jan  1 06:32:29.227
remote adjacency to GigabitEthernet0/0/0/0
Prefix Len 21, traffic index 0, precedence n/a, priority 15
Extensions: context-label:16001
   via 23.0.0.2/32, GigabitEthernet0/0/0/0, 12 dependencies, weight 0, class 0
[flags 0x0]
    path-idx 0 NHID 0x0 [0x1100a290 0x0]
    next hop 23.0.0.2/32
    remote adjacency
    local label 24001      labels imposed {16001}
RP/0/RP0/CPU0:P3#
```

The following output displays flags that indicate the merge operation:

```
RP/0/RP0/CPU0:P3#show cef 1.1.1.1/32 flags brief
1.1.1.1/32, version 320, labeled SR, internal 0x1000001 0x8310 (ptr 0xe831148)
[1], 0x600 (0xdfe4268), 0xa28 (0xebfe9b8)
leaf flags: owner locked, inserted
leaf flags2: LDP/SR merge requested,RIB pref over LSD,sr-pfx
leaf ext flags: PriChange,EXTERNAL_REACH_LC,L2TPV3_SPAN_DIAG_IFH_ENABLE,NH_
SEP_ILL,
Updated Jan  1 06:29:12.815
remote adjacency to GigabitEthernet0/0/0/0
Prefix Len 32, traffic index 0, precedence n/a, priority 1
   via 23.0.0.2/32, GigabitEthernet0/0/0/0, 12 dependencies, weight 0, class 0
[flags 0x0]
    path-idx 0 NHID 0x0 [0x1100a290 0x0]
    next hop 23.0.0.2/32
    remote adjacency
    local label 16001      labels imposed {16001}
RP/0/RP0/CPU0:P3#
```

The LDP to SR stitching was straightforward and did not require any intervention.

The next section will explore the SR to LDP LSP stitching for the reverse direction from PE1 to PE5.

SR to LDP stitching

LDP automatically assigns labels to routes upon their addition to the routing table. Consequently, an LDP label is generated for a specific route on every node within the LDP domain, irrespective of whether a distant router actively utilizes LDP.

In contrast, the allocation and distribution of labels within **Segment Routing** (**SR**) follow a more controlled process. Routers operating in SR-MPLS install all the Node-SIDs but restrict the inclusion of local Adjacency-SIDs in their **Label Forwarding Information Base** (**LFIB**). While LDP label distribution is unsolicited, SR-MPLS label allocation and distribution are somewhat solicited.

When a router is outside the SR-MPLS domain, it lacks a Prefix-sid to propagate in the IGP domain. Consequently, routers within the SR-MPLS domain lack an end-to-end label switch path to it.

The solution to this challenge will be explored in the next section.

Segment Routing Mapping Server (SRMS)

To establish an end-to-end LSP from the SR-MPLS to LDP label distribution domains, a **Segment Routing Mapping Server** (**SRMS**) is essential within the SR-MPLS domain. This server allocates and advertises Node-SIDs for non-SR routers, enabling SR-MPLS routers to install an outgoing SR-MPLS label in their LFIB.

Any router in the SR-MPLS domain can be configured to serve as an SRMS. In production networks, there might be more than one SRMS, for resilience.

The SRMS establishes a mapping that links Prefix-SIDs with corresponding IP addresses, uniquely for routers lacking SR capabilities. The IS-IS protocol advertises this mapping across the network, enabling routers in the SR domain to allocate and install SR-MPLS Prefix-sid labels for these non-SR routers.

The border routers between the SR-MPLS and LDP domains also learn Prefix-SIDs for non-SR routers and allocate corresponding labels. This enables the source SR-MPLS router to forward traffic along the SR-MPLS LSP until it reaches the border router. At this point, the incoming SR-MPLS label is merged with an outgoing LDP label to establish an end-to-end MPLS LSP.

In simple terms, the SRMS fulfills the SR-MPLS control-plane function for the non-SR routers, facilitating the advertisement of information about Prefix-sid labels that these routers cannot achieve independently, as elaborated on in the following section.

Configuration

As per the recommended approach in this book, the router chosen to act as an SRMS should be a router already operating within the SR network. In this context, the router selected for this role is P6.

P6

```
segment-routing
mapping-server
  prefix-sid-map
    address-family ipv4
      4.4.4.4/32 4
      5.5.5.5/32 5
      8.8.8.8/32 8
```

To implement this approach, the router P6, functioning as the SRMS, needs specific configurations, as shown, where the loopback IP address of P4, 4.4.4.4/32, is mapped with a Prefix-SID index of 4, and so on for P8 and PE5.

The following configuration ensures the seamless distribution of SRMS mapping information by IS-IS through its **Link State Protocol Data Units (LSPDUs)**:

```
router isis IGP
address-family ipv4 unicast
  segment-routing prefix-sid-map advertise-local
```

The segment-routing prefix-sid-map advertise-local command signals that the router is actively sharing its locally defined Prefix-SID mapping information with other routers in the network.

By default, all SR-MPLS routers are configured in receive mode, functioning as clients of the SRMS. Please be aware that the SRMS mapping can also be set up to include a series of labels assigned to a range of IP addresses within a subnet. However, this specific configuration is not discussed in this book.

The behavior of the SRMS is detailed in RFC 8661, titled **Segment Routing MPLS Interworking with LDP**.

The next section verifies that the preceding configuration has had the desired effect in the network.

Verification

In this section, we will confirm that SRMS has propagated the segment information for non-SR destinations. *Figure 4.4* illustrates the propagation of SRMS information, carrying the Prefix-SID index of non-SR routers across the SR-MPLS domain.

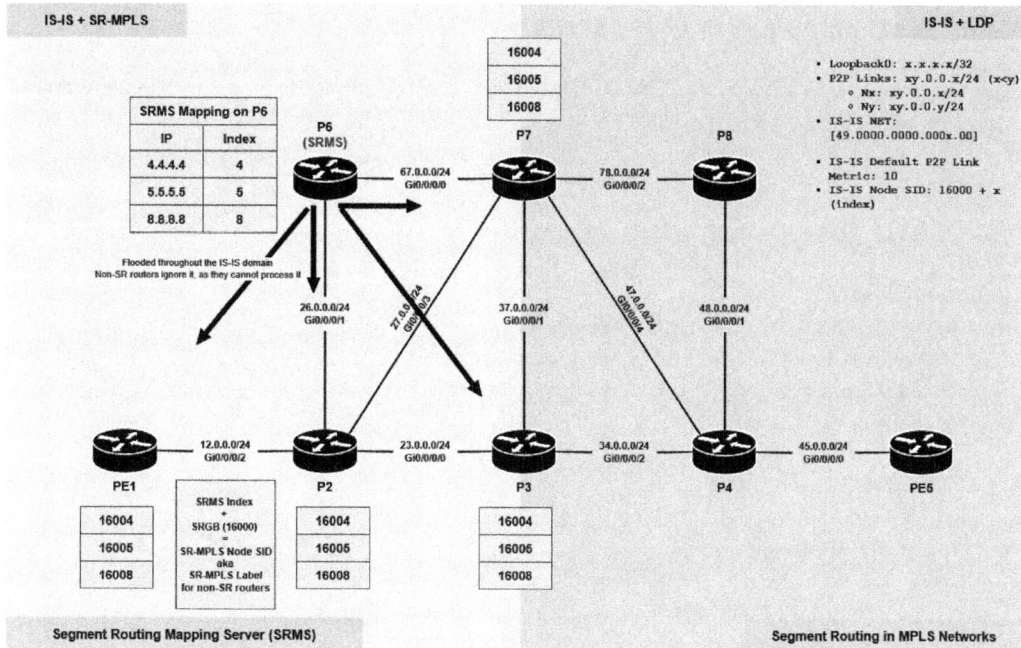

Figure 4.4 – Segment Routing Mapping Server (SRMS) operation

Verify the SRMS operation

The displayed output illustrates the mappings configured for the SRMS.

The selection of the range as 1 is deliberate, as it signifies a one-to-one mapping between prefixes and SIDs. This book does not explore the use of SID ranges for subnets:

```
RP/0/RP0/CPU0:P6#show segment-routing mapping-server prefix-sid-map ipv4
Prefix              SID Index    Range          Flags
4.4.4.4/32          4            1
5.5.5.5/32          5            1
8.8.8.8/32          8            1

Number of mapping entries: 3
RP/0/RP0/CPU0:P6
```

The mappings configured on the SRMS are then propagated throughout the IS-IS domain, as detailed in the next section. Non-SR routers, unable to interpret this information, will simply ignore it.

Verify SRMS mappings in IS-IS LSPDU

Upon inspecting P6's **Link State Protocol Data Unit (LSPDU)** on PE1, it becomes evident that the mappings are being advertised using TLV 149, as shown:

```
RP/0/RP0/CPU0:PE1#show isis database verbose internal P6


IS-IS IGP (Level-2) Link State Database
LSPID                   LSP Seq Num  LSP Checksum  LSP Holdtime/Rcvd  ATT/P/
OL  LSP Length
P6.00-00                0x0000000d   0xf03a        1150 /1200         0/0/0    279
  TLV code:1 length:2
    Area Address:    49
  TLV code:129 length:1
    NLPID:          0xcc
  TLV code:132 length:4
    IP Address:     6.6.6.6
  TLV code:135 length:45
    Metric: 0            IP-Extended 6.6.6.6/32
      SubTLV code:3 length:6
        Prefix-SID Index: 6, Algorithm:0, R:0 N:1 P:0 E:0 V:0 L:0
      SubTLV code:4 length:1
        Prefix Attribute Flags: X:0 R:0 N:1 E:0 A:0
    Metric: 10           IP-Extended 26.0.0.0/24
      SubTLV code:4 length:1
        Prefix Attribute Flags: X:0 R:0 N:0 E:0 A:0
    Metric: 10           IP-Extended 67.0.0.0/24
      SubTLV code:4 length:1
        Prefix Attribute Flags: X:0 R:0 N:0 E:0 A:0
  TLV code:137 length:2
    Hostname:       P6
  TLV code:22 length:92
    Metric: 10         IS-Extended P2.00
      SubTLV code:4 length:8
        Local Interface ID: 8, Remote Interface ID: 8
      SubTLV code:6 length:4
        Interface IP Address: 26.0.0.6
      SubTLV code:8 length:4
        Neighbor IP Address: 26.0.0.2
      SubTLV code:9 length:4
        Physical BW: 1000000 kbits/sec
      SubTLV code:31 length:5
        ADJ-SID: F:0 B:0 V:1 L:1 S:0 P:0 weight:0 Adjacency-sid:24021
    Metric: 10         IS-Extended P7.00
      SubTLV code:4 length:8
        Local Interface ID: 7, Remote Interface ID: 7
```

```
   SubTLV code:6 length:4
     Interface IP Address: 67.0.0.6
   SubTLV code:8 length:4
     Neighbor IP Address: 67.0.0.7
   SubTLV code:9 length:4
     Physical BW: 1000000 kbits/sec
   SubTLV code:31 length:5
     ADJ-SID: F:0 B:0 V:1 L:1 S:0 P:0 weight:0 Adjacency-sid:24023
```

The highlighted section here confirms that the SRMS information has been propagated to all routers in the IS-IS domain using TLV 149:

```
TLV code:149 length:17
  SID Binding:     4.4.4.4/32 F:0 M:0 S:0 D:0 A:0 Weight:0 Range:1
    SubTLV code:3 length:6
      SID: Start:4, Algorithm:0, R:0 N:0 P:0 E:0 V:0 L:0
TLV code:149 length:17
  SID Binding:     5.5.5.5/32 F:0 M:0 S:0 D:0 A:0 Weight:0 Range:1
    SubTLV code:3 length:6
      SID: Start:5, Algorithm:0, R:0 N:0 P:0 E:0 V:0 L:0
TLV code:149 length:17
  SID Binding:     8.8.8.8/32 F:0 M:0 S:0 D:0 A:0 Weight:0 Range:1
    SubTLV code:3 length:6
      SID: Start:8, Algorithm:0, R:0 N:0 P:0 E:0 V:0 L:0
TLV code:242 length:35
  Router Cap:     6.6.6.6 D:0 S:0
    SubTLV code:2 length:9
      Segment Routing: I:1 V:0, SRGB Base: 16000 Range: 8000
    SubTLV code:22 length:9
      SR Local Block: Base: 15000 Range: 1000
    SubTLV code:23 length:2
      Node Maximum SID Depth:
        Label Imposition: 10
    SubTLV code:19 length:2
      SR Algorithm:
        Algorithm: 0
        Algorithm: 1

Total Level-2 LSP count: 1    Local Level-2 LSP count: 0
RP/0/RP0/CPU0:PE1#
```

The details of the propagation of SR mappings through TLV 149 are explained here:

- TLV Code: 149: This is the SID/label-binding TLV, which is used to advertise prefixes to SID/label mappings by the SRMS, as described next. It holds the prefix for which the SID mapping is being shared. In the case of a one-to-one mapping, the range is set to 1:

  - F: **Address-Family Flag**. If unset, then the prefix carries an IPv4 prefix. If set, then the prefix carries an IPv6 prefix.

  - M: **Mirror Context Flag**. Set if the advertised SID corresponds to a mirrored context. The use of a mirrored context is described in RFC8402.

  - S: If set, the SID/label-binding TLV *should* be flooded across the entire routing domain. If the S-flag is not set, the SID/label-binding TLV *must not* be leaked between levels. This bit *must not* be altered during the TLV leaking.

  - D: When the SID/label-binding TLV is leaked from level 2 to level 1, the D-flag *must* be set. Otherwise, this flag *must* be clear. SID/label-binding TLVs with the D-Flag set *must not* be leaked from level 1 to level 2. This is to prevent TLV looping across levels.

  - A: **Attached Flag**. The originator of the SID/label-binding TLV *may* set the A bit in order to signal that the prefixes and SIDs advertised in the SID/label-binding TLV are directly connected to their originators. The mechanisms through which the originator of the SID/label-binding TLV can figure out whether a prefix is attached or not are outside the scope of this document (e.g., through explicit configuration). If the binding TLV is leaked to other areas/levels, the A-flag *must* be cleared.

  - SubTLV Code: 3: The SID remains consistent with what was observed earlier, in the Prefix-SID SubTLV, identified as SubTLV code: 3 within TLV code: 135 in *Chapter 3*. The LSPDU of P6, as examined on router PE1, confirms that the SRMS is advertising the Prefix-sid mappings as expected. The allocation of those mappings as labels is verified next.

Verify SRMS SIDs installed as labels

Upon receiving the Prefix-SID mapping from the SRMS, all SR routers have incorporated these mappings into their MPLS forwarding tables.

Despite not running SR-MPLS, P4, P8, and PE5 now have an SR-MPLS label associated with it on PE1, as shown:

```
RP/0/RP0/CPU0:PE1#show isis segment-routing label table

IS-IS IGP IS Label Table
Label           Prefix                   Interface
----------      ----------------         ---------
16001           1.1.1.1/32               Loopback0
16002           2.2.2.2/32
16003           3.3.3.3/32
16004           4.4.4.4/32
16005           5.5.5.5/32
16006           6.6.6.6/32
16007           7.7.7.7/32
16008           8.8.8.8/32
RP/0/RP0/CPU0:PE1#
```

Verify that the Prefix-SID index label associated with PE5, learned by the SRMS, is effectively installed in the LFIB of PE1:

```
RP/0/RP0/CPU0:PE1#show mpls forwarding prefix 5.5.5.5/32
Local  Outgoing    Prefix              Outgoing      Next Hop          Bytes
Label  Label       or ID               Interface                       Switched
------ ----------- ------------------- ------------- ---------------- ----------
--
16005  16005       SR Pfx (idx 5)      Gi0/0/0/2     12.0.0.2          0
RP/0/RP0/CPU0:PE1#
```

Even without originating the Prefix-sid label at source PE5, it is successfully programmed in the LFIB of the PE1 router, thanks to SRMS.

It should now enable the SR to LDP LSP stitch on the border router, as verified in the next section.

Verify SR to LDP label stitch on the border router

Once again, the responsibility for stitching the LSP falls upon the border routers. In this instance, the examination is focused solely on router P3 for convenience, as shown in *Figure 4.5*.

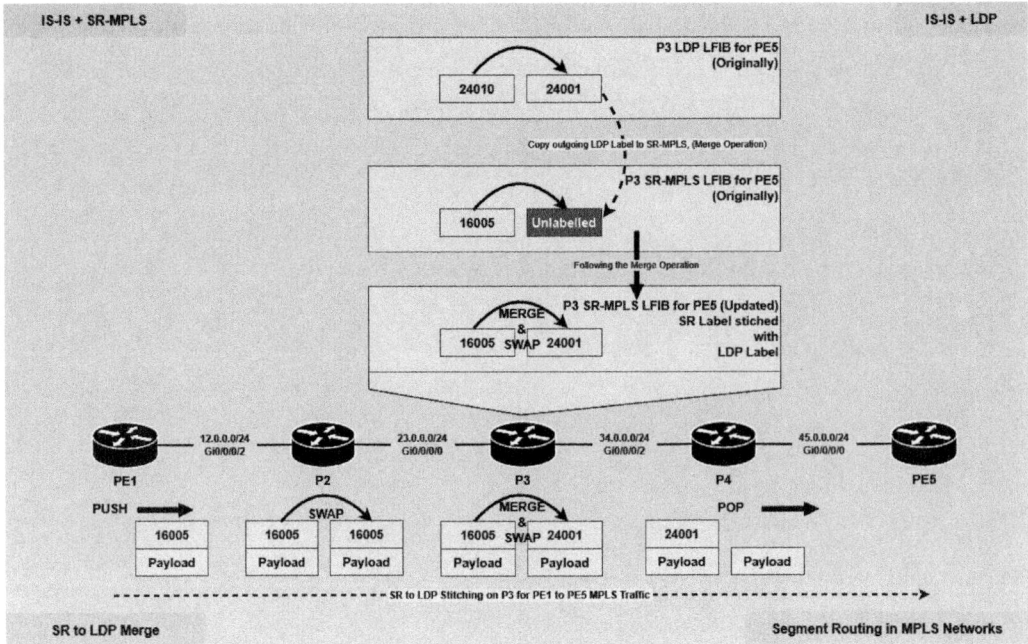

Figure 4.5 – SR to LDP merge operation

Router P3's designated next hop to reach PE5 is P4. However, P4 utilizes only LDP and does not support SR-MPLS, which leads to an outgoing LDP label without an associated SR-MPLS label, as illustrated:

```
RP/0/RP0/CPU0:P3#show mpls ldp forwarding 5.5.5.5/32


Codes:
   - = GR label recovering, (!) = LFA FRR pure backup path
   {} = Label stack with multi-line output for a routing path
   G = GR, S = Stale, R = Remote LFA FRR backup
   E = Entropy label capability

Prefix          Label   Label(s)      Outgoing      Next Hop             Flags
                In      Out           Interface                          G S R
E
--------------- ------- ------------- ------------- -------------------- -----
--
5.5.5.5/32      24010   24001         Gi0/0/0/2     34.0.0.4

RP/0/RP0/CPU0:P3#
```

Thus, the router replicates the outbound LDP label to the corresponding local SR-MPLS label designated for PE5. This process, referred to as a merge operation, occurs rapidly, making it challenging to observe in real-time:

```
RP/0/RP0/CPU0:P3#show mpls forwarding prefix 5.5.5.5/32
Local  Outgoing    Prefix             Outgoing      Next Hop         Bytes
Label  Label       or ID              Interface                      Switched
------ ----------- ------------------ ------------- ---------------- ----------
--
16005  24001       SR Pfx (idx 5)     Gi0/0/0/2     34.0.0.4         0
RP/0/RP0/CPU0:P3#
```

Due to the consistent SRGB offset of 16000 across the SR domain, the SRMS label index generated a consistent SR-MPLS label for non-SR routers, such as 16005 for PE5.

The SR-MPLS process on the border router P3 used this information, learned from the SRMS router P6, to create the SR-MPLS label for PE5. The outgoing router from P3 to PE5 is P4, which runs only LDP and has distributed an LDP label for PE5 to P3.

Router P3 merges the local SR-MPLS label with the outgoing LDP label to stitch the end-to-end MPLS LSP, thereby completing the merge operation.

The merge operation is also evident in the FIB, thanks to the inclusion of the `sr-prefer` configuration within the router's IS-IS configuration.

The flags displayed in the following output serve as conclusive evidence of the completed merge operation:

```
RP/0/RP0/CPU0:P3#show cef 5.5.5.5/32 flags brief
5.5.5.5/32, version 345, labeled SR, internal 0x1000001 0x87f0 (ptr 0xe82fae0)
[1], 0x600 (0xdfe2b10), 0xa28 (0xebfea08)
leaf flags: owner locked, inserted
leaf flags2: LDP/SR merge requested,RIB pref over LSD,LDP/SR merge ac-
tive,sr-pfx
leaf ext flags: PriChange,EXTERNAL_REACH_LC,L2TPV3_SPAN_DIAG_IFH_ENABLE,NH_
SEP_ILL,
Updated Jan  1 06:48:43.119
remote adjacency to GigabitEthernet0/0/0/2
Prefix Len 32, traffic index 0, precedence n/a, priority 15
   via 34.0.0.4/32, GigabitEthernet0/0/0/2, 16 dependencies, weight 0, class 0
[flags 0x0]
    path-idx 0 NHID 0x0 [0x1100a1f0 0x0]
    next hop 34.0.0.4/32
    remote adjacency
    local label 16005      labels imposed {24001}
RP/0/RP0/CPU0:P3#
```

The traceroute verifies the merge operation in the data plane in the following section.

Verify SR-MPLS traceroute

With the SR-MPLS label smoothly stitched into the LDP label, let's explore the entire LSP from PE1 (utilizing SR-MPLS) to PE5 (using LDP).

The seamless label stitching process equally applies to both ECMP routes:

```
RP/0/RP0/CPU0:PE1#traceroute sr-mpls multipath 5.5.5.5/32 verbose

Starting LSP Path Discovery for 5.5.5.5/32

Codes: '!' - success, 'Q' - request not sent, '.' - timeout,
  'L' - labeled output interface, 'B' - unlabeled output interface,
  'D' - DS Map mismatch, 'F' - no FEC mapping, 'f' - FEC mismatch,
  'M' - malformed request, 'm' - unsupported tlvs, 'N' - no rx label,
  'P' - no rx intf label prot, 'p' - premature termination of LSP,
  'R' - transit router, 'I' - unknown upstream index,
  'X' - unknown return code, 'x' - return code 0

Type escape sequence to abort.

LLL!
Path 0 found,
output interface GigabitEthernet0/0/0/2 nexthop 12.0.0.2
source 12.0.0.1 destination 127.0.0.1
  0 12.0.0.1 12.0.0.2 MRU 1500 [Labels: 16005 Exp: 0] multipaths 0
L 1 12.0.0.2 23.0.0.3 MRU 1500 [Labels: 16005 Exp: 0] ret code 8 multipaths 2
L 2 23.0.0.3 34.0.0.4 MRU 1500 [Labels: 24001 Exp: 0] ret code 8 multipaths 1
L 3 34.0.0.4 45.0.0.5 MRU 1500 [Labels: implicit-null Exp: 0] ret code 8
multipaths 1
! 4 45.0.0.5, ret code 3 multipaths 0
```

As requested by the traceroute command, the multipath keyword has traced all paths. The following output describes an additional path:

```
LL!
Path 1 found,
output interface GigabitEthernet0/0/0/2 nexthop 12.0.0.2
source 12.0.0.1 destination 127.0.0.0
  0 12.0.0.1 12.0.0.2 MRU 1500 [Labels: 16005 Exp: 0] multipaths 0
L 1 12.0.0.2 27.0.0.7 MRU 1500 [Labels: 16005 Exp: 0] ret code 8 multipaths 2
L 2 27.0.0.7 47.0.0.4 MRU 1500 [Labels: 24001 Exp: 0] ret code 8 multipaths 1
L 3 47.0.0.4 45.0.0.5 MRU 1500 [Labels: implicit-null Exp: 0] ret code 8
multipaths 1
! 4 45.0.0.5, ret code 3 multipaths 0

Paths (found/broken/unexplored) (2/0/0)
```

```
Echo Request (sent/fail) (7/0)
Echo Reply (received/timeout) (7/0)
Total Time Elapsed 118 ms
RP/0/RP0/CPU0:PE1#
```

The aforementioned output confirms the seamless stitching of the MPLS LSP from the SR-MPLS domain to the LDP domain.

The diagram in *Figure 4.6* illustrates the traceroute operation described.

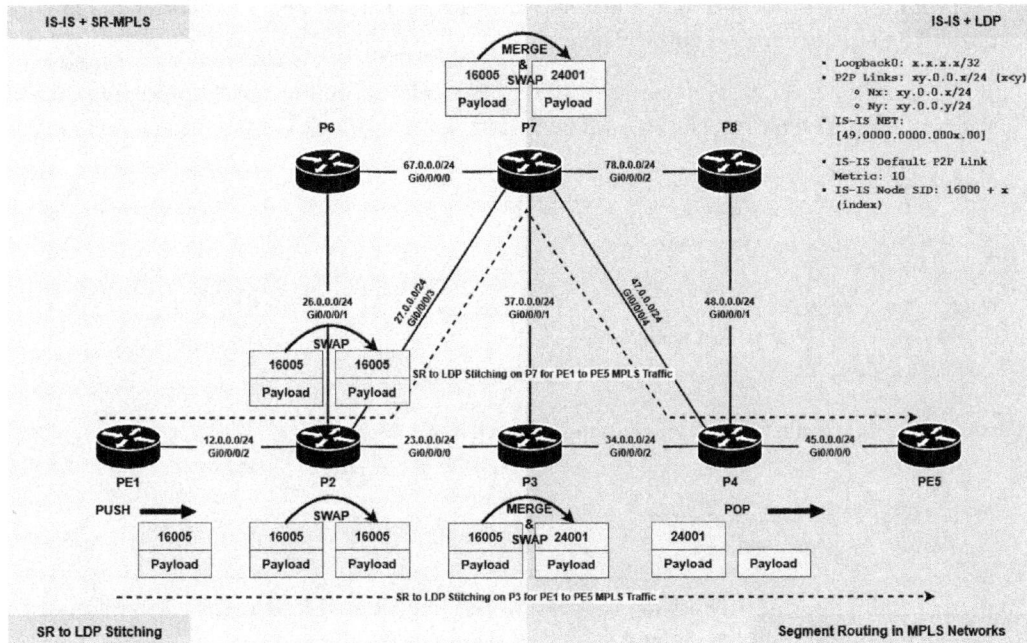

Figure 4.6 – SR to LDP stitching

To facilitate this stitching, the presence of an SRMS is essential.

> **Important**
>
> In the context of SR-LDP Interworking on the Cisco IOS-XR platform, the label stitching operation occurs automatically for both LDP to SR and SR to LDP. When a local label is distributed by one protocol but lacks a corresponding outgoing label from the same protocol (while having one from a different protocol), the local label seamlessly merges with the outgoing label in the LFIB and the FIB. This integration ensures the establishment of an end-to-end label-switched path.

This concludes the SR-LDP interworking lab. In the upcoming chapters, it is crucial that the entire topology operates exclusively on SR-MPLS. The next section will concentrate on migrating the remaining LDP routers to the SR-MPLS domain.

Migrating to SR-MPLS

The migration is facilitated by extending the SR-MPLS configuration to the LDP routers and removing the LDP configuration. In production networks, migrating to `sr-prefer` should be carried out during a planned window as a precaution.

Configuration

The remaining routers not operating under SR-MPLS are P4, P8, and PE5. Extend the configuration of SR-MPLS to these routers as outlined here:

In this context, the x variable represents the IDs of the nodes, as outlined in the preceding chapter. Specifically, x equals 4 for P4, 5 for PE5, and 8 for P8:

```
router isis IGP
address-family ipv4 unicast
  segment-routing mpls sr-prefer
!
interface Loopback0
  address-family ipv4 unicast
   prefix-sid index x
```

Furthermore, deactivate LDP on those routers, as well as on P3 and P7:

```
router isis IGP
address-family ipv4 unicast
  no mpls ldp auto-config
```

At the end of this step, the network topology is now running SR-MPLS exclusively. This means there is no need for SRMS to advertise Prefix-SIDs, so it is removed in the next section.

Stopping SRMS advertisements

Since all the routers are now advertising their Prefix-SID labels independently, the SRMS advertisements can be stopped by applying the following configuration.

```
router isis IGP
address-family ipv4 unicast
  no segment-routing prefix-sid-map advertise-local
 !
 !
no segment-routing
```

> **Tip**
>
> On P3, customize the default SRGB by changing it from 16000-23999 to a new contiguous range, such as 50000-60000, using the `segment-routing global-block 50000 60000` command. Subsequently, IS-IS will request the new segment range from LSD for label encoding, triggering a network-wide update about the updated SRGB range. P3 will withdraw local labels from the previous 16000-23999 range and initiate label allocations from the new 50000-60000 segment range. This process causes a brief disruption in labeled traffic flow.
>
> To observe the impact on label allocation, distribution, and operation, re-verify the RIB, FIB, and LFIB commands, along with traceroute. This process allows for a more focused examination of the changes and aids in enhancing your understanding of labeled traffic dynamics within the network.
>
> While not mandatory, for clarity and ease of understanding, please revert the changes before proceeding to the next lab by using the `no segment-routing global-block` command.

This marks the completion of the SR-LDP interworking and the migration of the entire network exclusively to the SR-MPLS domain.

> **Note**
>
> Bear in mind that the Prefix-SIDs obtained through standard IS-IS advertisements take precedence over those learned from the SRMS. Regardless, to ensure a clean configuration, discontinue the advertising of SRMS mapping into IS-IS on P6.

Summary

This chapter explored the communication dynamics within a network split into two distinct MPLS data planes: one running exclusively on SR-MPLS and the other solely on LDP. This transition involved removing LDP from one portion of the network, where SR-MPLS was previously introduced, effectively transitioning it to an SR-only environment.

To validate this transition, the lab investigated how **Label Switched Paths** (**LSPs**) transition between LDP and SR. It's noted that the IOS-XR platform seamlessly handles the transition from LDP to SR LSPs. However, for SR to LDP LSPs, an additional component called the **Segment Routing Mapping Server** (**SRMS**) is necessary. The SRMS plays a vital role in generating and distributing SR labels or SIDs within the SR-MPLS network for non-SR routers, enabling the establishment of LSPs across SR and LDP domains.

Throughout the validation process, the lab ensures the smooth operation of LSPs across various MPLS domains in both directions.

The chapter concluded with the successful migration of the entire network topology to an SR-MPLS-only configuration. This comprehensive transition represents a significant milestone in the evolution of MPLS networks.

After migrating the network to an SR-MPLS-only setup, the next chapters will explore **Topology Independent Loop-Free Alternate (TI-LFA)** link protection fast reroute scenarios.

References

Below are the key references cited throughout this chapter. These documents provide additional insights and technical details on MPLS, Fast Reroute, and Segment Routing, offering further reading for those interested in exploring these topics in depth.

- *[RFC7855] Previdi, S., Ed., Filsfils, C., Ed., Decraene, B., Litkowski, S., Horneffer, M., and R. Shakir, "Source Packet Routing in Networking (SPRING) Problem Statement and Requirements", RFC 7855, DOI 10.17487/RFC7855, May 2016, <https://www.rfc-editor.org/info/rfc7855>.*

- *[RFC8661] Bashandy, A., Ed., Filsfils, C., Ed., Previdi, S., Decraene, B., and S. Litkowski, "Segment Routing MPLS Interworking with LDP", RFC 8661, DOI 10.17487/RFC8661, December 2019, <https://www.rfc-editor.org/info/rfc8661>.*

Part 3 -
Fast Reroute in
SR-MPLS Networks

This part explores the **Fast Reroute** (**FRR**) mechanisms in SR-MPLS networks, specifically focusing on how traffic can be efficiently rerouted using TI-LFA. It also provides comprehensive coverage of multiple TI-LFA scenarios.

This part contains the following chapters:

Lab 4 – Introducing TI-LFA (Topology Independent – Loop-Free Alternate)

Modern networks handle a substantial volume and variety of traffic, including data, voice, video, online gaming, and the **IoT** (**Internet of Things**). When a link or a router fails, the network experiences a convergence delay. This delay involves detecting the fault, updating the routers, calculating and installing an alternative path to restore connectivity, and then allowing applications to re-establish connections and resume communication. This brief outage often exceeds tolerable limits, disrupting the user experience.

To mitigate these interruptions, fast reroute mechanisms have been developed.

The widely used fast reroute method in segment routing networks is **TI-LFA** (**Topology Independent Loop-Free Alternate**). TI-LFA Fast Reroute protection enhances network resilience by creating backup routes for each entry in the **Interior Gateway Protocol** (**IGP**) routing table or the **Routing Information Base** (**RIB**) whenever available. It adheres to the same principles of segment routing, where the entire path is encoded within the packet as a label stack. The TI-LFA fast reroute path is similarly encoded as a label stack.

Objective

The objective of this chapter is to introduce TI-LFA in the SR-MPLS network topology in *Figure 5.1*. It will involve verifying the availability of backup paths and understanding how they are calculated and installed in the routers. These goals are accomplished through the following tasks:

1. Configuring TI-LFA on all IS-IS connections for every router.
2. Verifying that TI-LFA is operationally enabled.

3. Verifying that the TI-LFA backup route is computed and added to the RIB, FIB, and LFIB.

4. Understanding the methods to calculate backup paths.

TI-LFA will be enabled on all routers within the SR-MPLS network topology, illustrated in *Figure 5.1*.

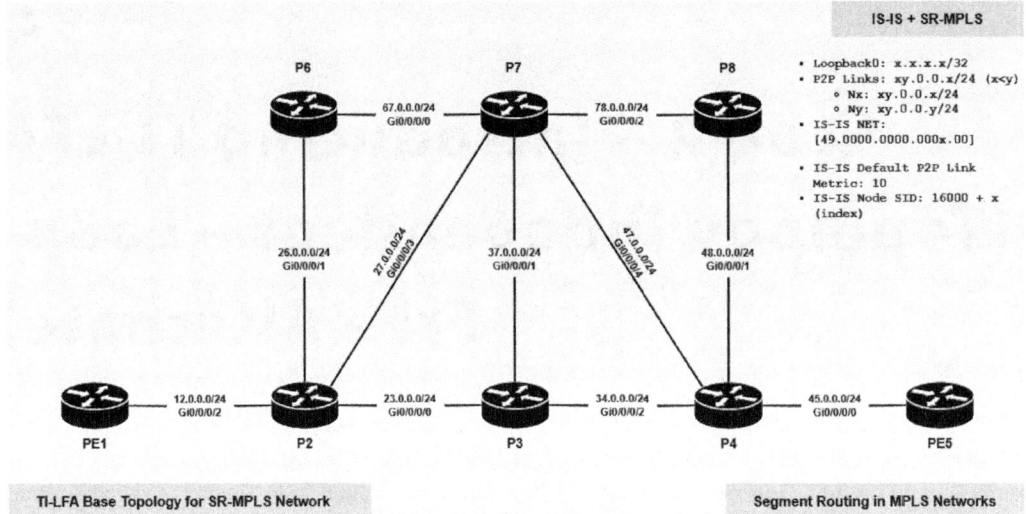

Figure 5.1 – The TI-LFA base topology

> **Note**
>
> For all upcoming labs in the book, P2 will serve as the **Point of Local Repair** (**PLR**) due to its resilient and redundant multiple paths in the core, ensuring robust forwarding of traffic from PE1 to PE5.

Before diving into the lab work, let's first take a moment to understand the significance of resiliency and fast rerouting in production networks.

Background of fast rerouting

Network incidents, such as link failures and node failures, can disrupt the seamless flow of data within a network. Network resiliency is the ability of a network to withstand and recover from such disruptions. When incidents occur, IGP convergence comes into play, recalculating the routing table to find alternative paths through resilient resources. However, during this convergence period, there's a risk of traffic blackholes, where data may get lost or delayed. Hence, it's crucial to have the ability to forward traffic even during convergence to ensure data continuity and a seamless user experience.

Traditional MPLS networks, using RSVP-TE, install a pre-established backup **Label Switch Path** (**LSP**) to quickly reroute traffic if there are network failures. However, this approach increases the load on routers, as the backup LSPs function as additional RSVP-TE sessions, requiring routers to manage more sessions. When a failure occurs, the router switches the traffic to the backup LSP while the IGP converges. After a fixed wait time, the router then tries to re-establish the primary LSP, causing additional processing and memory usage on the routers.

Segment routing doesn't maintain sessions or states between routers. As showcased in the preceding chapters, the label-switched path for SR-MPLS relies entirely on the IGP. The TI-LFA fast-rerouting method within the SR-MPLS network also relies on the IGP and doesn't retain any state.

The journey to **Topology Independent Loop-Free Alternate** (**TI-LFA**) represents an evolution of IP **Fast Reroute** (**FRR**) mechanisms, addressing the limitations of earlier approaches. Here is a summary of this progression.

Classic LFA

Classic LFA, or just LFA, is a fast reroute mechanism designed to enhance the resilience of IP networks by providing an alternate path for traffic when a primary link fails. It was first defined in RFC 5286, titled "*Basic Specification for IP Fast Reroute: Loop-Free Alternates.*" As part of the IP FRR framework, it offers sub-50 millisecond recovery times, making it suitable for maintaining high network availability. The primary objective of LFA is to precompute backup paths that can be immediately utilized when a link or node failure occurs, minimizing packet loss and maintaining service continuity.

The key concepts are as follows:

- **Backup path calculation**: LFA precomputes alternate paths that avoid the failed component (link or node) and ensures that these paths do not form a loop.

- **Inequality condition**: For an alternate path to be valid, it must satisfy the inequality condition, ensuring that the backup path does not loop back to the failure point. As per RFC 5286, **Inequality 1: Loop-Free Criterion** is defined as, "*A neighbor N can provide an LFA if and only if* `Distance_opt(N, D) < Distance_opt(N, S) + Distance_opt(S, D.`"

   Here, S is used to indicate the calculating router. `N_i` is a neighbor of S; N is used as an abbreviation when only one neighbor is being discussed. D is the destination under consideration.

- **Ease of deployment**: LFA can be easily integrated into existing network architectures without requiring significant changes to the infrastructure.

- **Incomplete coverage**: LFA is topology-dependent and doesn't provide protection for all destinations in all network topologies.

- **Suboptimal backup paths**: The backup paths may not always be the most efficient or aligned with network planning.

Despite its advantages, classic LFA has limitations, particularly in networks with complex topologies, and finding feasible alternate paths can be challenging.

Remote LFA

Remote LFA (RLFA) was first defined in RFC 7490 titled, "*Remote Loop-Free Alternate (LFA) Fast Reroute (FRR),*" extending the capabilities of classic LFA by addressing scenarios where local alternates are inadequate. RLFA allows the use of remote nodes as backup next hops, thereby increasing the probability of finding a loop-free alternate path in complex network topologies. The objective is to pre-establish a repair tunnel toward an alternate next-hop that is free from loops and accessible if there is a link or node failure. However, it is important to note that this method is also topology-dependent, and in certain scenarios, these backup paths may be inaccessible due to the metrics associated with the network links.

The key concepts are as follows:

- **PQ-Space concept**: Remote LFA introduces the PQ-Space concept, where PQ represents the remote node used to reach the destination via an alternate path.

- **Tunneling**: Traffic is tunneled to the remote node "PQ" using MPLS encapsulation.

- **Extended coverage**: By leveraging remote nodes, RLFA enhances the likelihood of finding feasible alternate paths, thereby improving network resilience.

RLFA, like classic LFA, is topology-dependent, which means it may not provide 100% backup path coverage in complex network topologies.

Additionally, RFC 7490 defines terminologies that are essential to understand TI-LFA:

- **Repair tunnel**: A tunnel established for the purpose of providing a virtual neighbor that is an LFA.

- **P-Space**: The P-Space of a router with respect to a protected link is the set of routers reachable from that specific router using the pre-convergence shortest paths, without any of those paths (including equal-cost path splits) transiting that protected link.

- **Extended P-Space**: Consider the set of neighbors of a router protecting a link. Exclude from that set of routers the router reachable over the protected link. The extended P-Space of the protecting router with respect to the protected link is the union of the P-Spaces of the neighbors in that set of neighbors with respect to the protected link.

- **Q-Space**: The Q-Space of a router with respect to a protected link is the set of routers from which that specific router can be reached, without any path (including equal-cost path splits) transiting that protected link.

- **PQ node**: A PQ node of a S node with respect to a protected link, S-E, is a node that is a member of both the P-Space (or the extended P-Space) of S, with respect to that protected link, and the Q-Space of E, with respect to that protected link. A repair tunnel endpoint is chosen from the set of PQ nodes.

- **RLFA**: The use of a PQ node rather than a neighbor of the repairing node as the next hop in an LFA repair [RFC5286].

In RLFA, the objective is to identify a PQ node (a node in both extended P-Space and Q-Space) within the topology. The aim is to establish a repair tunnel leading to that node, ensuring that traffic can then proceed to the destination without encountering any loops along the path if there is a failure.

In specific situations, when there is no overlap between the P- and Q- Space, RLFA does not offer complete backup path coverage.

Topology Independent Loop-Free Alternate (TI-LFA)

TI-LFA represents a significant advancement in FRR technologies, providing robust coverage across all network topologies, including those where classic and RLFA may have limitations. TI-LFA extends the RLFA algorithm by calculating a backup tunnel where the P and Q nodes are not the same. In TI-LFA, a repair tunnel is established to the nearest Q node along the post-convergence path. This repair tunnel leverages the shortest path to the P node and a source-routed path from the P node to the Q node. This ensures universal coverage without resorting to sub-optimal routing.

This approach results in achieving 100% backup path coverage in the network. `draft-ietf-rtgwg-segment-routing-ti-lfa`, also known as "**Topology Independent Fast Reroute using Segment Routing**," outlines the methods used to identify TI-LFA backup paths for fast rerouting traffic in the event of a failure.

The key concepts are as follows:

- **100% backup path coverage**: TI-LFA ensures complete coverage in any multi-link network topology.

- **Segment Routing (SR)**: Utilizes SR to eliminate the need for additional protocols or sessions, simplifying network operations.

- **IGP Integration**: Utilizes the **Link State Database** (**LSDB**) of the IGP, avoiding the need for additional state or protocols.

- **Path alignment**: Aligns the backup path with the expected post-convergence path, optimizing network efficiency and avoiding suboptimal paths.

Thanks to segment routing, TI-LFA can offer a post-convergence path as a backup, encoded as a list of segments in a label stack. Without segment routing, the post-convergence path often cannot be used, as it is not loop-free in many cases.

The FRR calculation mechanism used in this book employs a per-prefix or per-destination approach, meaning that for each route in the routing table, a backup path is computed and installed in the RIB, LFIB, and FIB. Classic LFA and TI-LFA protections will be specifically enabled for per-prefix backup paths in the upcoming sections, where the importance of enabling classic LFA alongside TI-LFA will also be demonstrated.

In summary, while classic LFA offers fundamental FRR capabilities in simpler topologies and RLFA extends this to more complex scenarios, TI-LFA provides a topology-independent solution that ensures comprehensive coverage and streamlined operations through segment routing. Each of these approaches builds upon the previous ones, addressing their respective limitations and enhancing overall network resilience and efficiency.

In the upcoming chapters of the book, there will be a thorough examination of scenario-based calculations to identify P and Q nodes in a network. This aims to enhance your understanding by providing detailed insights into the practical application of the concepts.

Configuration

Since TI-LFA leverages the IGP for computing backup routes, its activation in the Cisco IOS-XR software involves applying two essential commands to the physical interfaces of the IS-IS link-state routing protocol.

The `fast-reroute per-prefix` command exclusively enables classic LFA. Subsequently, the `fast-reroute per-prefix ti-lfa` command, added after the previous one, activates TI-LFA.

Both commands are essential for enabling TI-LFA protection, and this process is shown further as follows:

- The PE1 router is connected only to P2, so TI-LFA is activated on that specific interface under the IS-IS configuration context:

    PE1

    ```
    router isis IGP
    interface GigabitEthernet0/0/0/2
      address-family ipv4 unicast
        fast-reroute per-prefix
        fast-reroute per-prefix ti-lfa
    ```

- On the P2 router, TI-LFA is activated on the interfaces connected to PE1, P3, P6, and P7, as shown here:

    P2

    ```
    router isis IGP
    interface GigabitEthernet0/0/0/0
      address-family ipv4 unicast
        fast-reroute per-prefix
        fast-reroute per-prefix ti-lfa
      !
    !
    interface GigabitEthernet0/0/0/1
      address-family ipv4 unicast
        fast-reroute per-prefix
        fast-reroute per-prefix ti-lfa
      !
    !
    interface GigabitEthernet0/0/0/2
      address-family ipv4 unicast
        fast-reroute per-prefix
    ```

```
    fast-reroute per-prefix ti-lfa
    !
  !
  interface GigabitEthernet0/0/0/3
    address-family ipv4 unicast
    fast-reroute per-prefix
    fast-reroute per-prefix ti-lfa
```

- On the P3 router, TI-LFA is activated on the interfaces connected to P2, P4, and P7, as shown here:

P3

```
  router isis IGP
  interface GigabitEthernet0/0/0/0
    address-family ipv4 unicast
    fast-reroute per-prefix
    fast-reroute per-prefix ti-lfa
    !
  !
  interface GigabitEthernet0/0/0/1
    address-family ipv4 unicast
    fast-reroute per-prefix
    fast-reroute per-prefix ti-lfa
    !
  !
  interface GigabitEthernet0/0/0/2
    address-family ipv4 unicast
    fast-reroute per-prefix
    fast-reroute per-prefix ti-lfa
```

- On the P4 router, TI-LFA is activated on the interfaces connected to P3, P8, and PE5, as shown here:

P4

```
  router isis IGP
  interface GigabitEthernet0/0/0/0
    address-family ipv4 unicast
    fast-reroute per-prefix
    fast-reroute per-prefix ti-lfa
    !
  !
  interface GigabitEthernet0/0/0/1
    address-family ipv4 unicast
    fast-reroute per-prefix
    fast-reroute per-prefix ti-lfa
    !
```

```
!
interface GigabitEthernet0/0/0/2
  address-family ipv4 unicast
    fast-reroute per-prefix
    fast-reroute per-prefix ti-lfa
  !
!
interface GigabitEthernet0/0/0/4
  address-family ipv4 unicast
    fast-reroute per-prefix
    fast-reroute per-prefix ti-lfa
```

- The PE5 router is connected only to P4, so TI-LFA is activated on that specific interface under the IS-IS configuration context:

PE5

```
router isis IGP
interface GigabitEthernet0/0/0/0
  address-family ipv4 unicast
    fast-reroute per-prefix
    fast-reroute per-prefix ti-lfa
```

- On the P6 router, TI-LFA is activated on the interfaces connected to P2 and P7, as shown here:

P6

```
router isis IGP
interface GigabitEthernet0/0/0/0
  address-family ipv4 unicast
    fast-reroute per-prefix
    fast-reroute per-prefix ti-lfa
  !
!
interface GigabitEthernet0/0/0/1
  address-family ipv4 unicast
    fast-reroute per-prefix
    fast-reroute per-prefix ti-lfa
```

- On the P7 router, TI-LFA is activated on the interfaces connected to P2, P3, P4, P6, and P8, as shown here:

P7

```
router isis IGP
interface GigabitEthernet0/0/0/0
  address-family ipv4 unicast
    fast-reroute per-prefix
```

```
    fast-reroute per-prefix ti-lfa
    !
  !
  interface GigabitEthernet0/0/0/1
    address-family ipv4 unicast
      fast-reroute per-prefix
      fast-reroute per-prefix ti-lfa
    !
  !
  interface GigabitEthernet0/0/0/2
    address-family ipv4 unicast
      fast-reroute per-prefix
      fast-reroute per-prefix ti-lfa
    !
  !
  interface GigabitEthernet0/0/0/3
    address-family ipv4 unicast
      fast-reroute per-prefix
      fast-reroute per-prefix ti-lfa
    !
  !
  interface GigabitEthernet0/0/0/4
    address-family ipv4 unicast
      fast-reroute per-prefix
      fast-reroute per-prefix ti-lfa
```

- On the P8 router, TI-LFA is activated on the interfaces connected to P4 and P7, as shown here:

P8

```
  router isis IGP
  interface GigabitEthernet0/0/0/1
    address-family ipv4 unicast
      fast-reroute per-prefix
      fast-reroute per-prefix ti-lfa
    !
  !
  interface GigabitEthernet0/0/0/2
    address-family ipv4 unicast
      fast-reroute per-prefix
      fast-reroute per-prefix ti-lfa
```

The preceding configuration activates TI-LFA, enabling the routers to calculate and install a backup route for every route in the routing table, as verified in the next section.

Verification

In this section, we will verify that the TI-LFA configuration is effective. For each primary path, ensure that, wherever available, a backup path has been calculated and installed in the RIB, LFIB, and FIB.

Verifying TI-LFA activation

The following output on the P2 router confirms the activation of TI-LFA on the interfaces involved in the IS-IS IGP:

```
RP/0/RP0/CPU0:P2#show isis interface Gi0/0/0/0

GigabitEthernet0/0/0/0        Enabled
  Adjacency Formation:        Enabled
  Prefix Advertisement:       Enabled
  IPv4 BFD:                   Disabled
  IPv6 BFD:                   Disabled
  BFD Min Interval:           150
  BFD Multiplier:             3
  RSI SRLG:                   Registered
  Bandwidth:                  1000000

  Circuit Type:               level-2-only
  Media Type:                 P2P
  Circuit Number:             0
  Measured Delay:             Min:- Avg:- Max:- usec
  Delay Normalization:        Interval:0 Offset:0
  Normalized Delay:           Min:- Avg:- Max:- usec
  Link Loss:                  -
  Extended Circuit Number:    7
  Next P2P IIH in:            7 s
  LSP Rexmit Queue Size:      0

  Level-2
    Adjacency Count:          1
    LSP Pacing Interval:      33 ms
    PSNP Entry Queue Size:    0
    Hello Interval:           10 s
    Hello Multiplier:         3

  CLNS I/O
    Protocol State:           Up
    MTU:                      1497
    SNPA:                     5005.0004.0003
    Layer-2 MCast Groups Membership:
```

```
      All ISs:              Yes

IPv4 Unicast Topology:     Enabled
  Adjacency Formation:     Running
  Prefix Advertisement:    Running
        Policy (L1/L2):    -/-
  Metric (L1/L2):          0/10
  Metric fallback:
    Bandwidth (L1/L2):     Inactive/Inactive
    Anomaly (L1/L2):       Inactive/Inactive
  Weight (L1/L2):          0/0
  MPLS Max Label Stack:    3/3/10/10 (PRI/BKP/SRTE/SRAT)
  MPLS LDP Sync (L1/L2):   Disabled/Disabled
  FRR (L1/L2):             L1 Enabled       L2 Enabled
    FRR Type:              per-prefix       per-prefix
    Direct LFA:            Enabled          Enabled
    Remote LFA:            Not Enabled      Not Enabled
     Tie Breaker           Default          Default
     Line-card disjoint    30               30
     Lowest backup metric  20               20
     Node protecting       40               40
     Primary path          10               10
    TI LFA:                Enabled          Enabled
     Tie Breaker           Default          Default
    Link Protecting        Enabled          Enabled
     Line-card disjoint    0                0
     Node protecting       0                0
     SRLG disjoint         0                0

IPv4 Address Family:       Enabled
  Protocol State:          Up
  Forwarding Address(es):  23.0.0.2
  Global Prefix(es):       23.0.0.0/24

LSP transmit timer expires in 0 ms
LSP transmission is idle
Can send up to 9 back-to-back LSPs in the next 0 ms

RP/0/RP0/CPU0:P2#
```

Additionally, it indicates that link protection is enabled within TI-LFA. It's important to note that link protection is the default method and cannot be disabled. This is because if other protection methods are configured but no backup route is accessible, the system automatically defaults to link protection. This specific scenario will be further explored later in the book.

A similar output is expected for all interfaces on all routers participating in IS-IS.

Verifying backup path availability

The following output confirms the existence of a pre-computed backup route for the primary route.

The following command on the P2 router checks the FRR path for the destination router, PE5, which has the loopback IP address 5.5.5.5/32:

```
RP/0/RP0/CPU0:P2#show isis fast-reroute detail 5.5.5.5/32


L2 5.5.5.5/32 [30/115] Label: 16005, medium priority
    Installed Jan 01 06:58:23.045 for 00:01:40
      via 23.0.0.3, GigabitEthernet0/0/0/0, Label: 16005, P3, SRGB Base: 16000,
Weight: 0
        Backup path: LFA, via 27.0.0.7, GigabitEthernet0/0/0/3, Label: 16005,
P7, SRGB Base: 16000, Weight: 0, Metric: 30
        P: Yes, TM: 30, LC: No, NP: Yes, D: Yes, SRLG: Yes
      via 27.0.0.7, GigabitEthernet0/0/0/3, Label: 16005, P7, SRGB Base: 16000,
Weight: 0
        Backup path: LFA, via 23.0.0.3, GigabitEthernet0/0/0/0, Label: 16005,
P3, SRGB Base: 16000, Weight: 0, Metric: 30
        P: Yes, TM: 30, LC: No, NP: Yes, D: Yes, SRLG: Yes
      src PE5.00-00, 5.5.5.5, prefix-SID index 5, R:0 N:1 P:0 E:0 V:0 L:0,
Alg:0
RP/0/RP0/CPU0:P2#
```

The details provided in the output have undergone changes in recent versions of the Cisco IOS XR software. Nevertheless, here's a guide to understanding the relevant elements:

- `L2 5.5.5.5/32 [30/115] Label: 16005, medium priority`: This line offers details about the IPv4 prefix, including its administrative distance (115), the end-to-end metric (30), and the associated label (16005). The presence of the label signifies MPLS forwarding information for this specific prefix.

- `via 23.0.0.3, GigabitEthernet0/0/0/0, Label: 16005, P3, SRGB Base: 16000, Weight: 0`: This describes the next-hop IP of the primary path as 23.0.0.3, traversing the GigabitEthernet0/0/0/0 interface. The next-hop router is identified as P3, and its **Segment Routing Global Block** (**SRGB**) offset is identified at 16000.

- `Backup path: LFA, via 27.0.0.7, GigabitEthernet0/0/0/3, Label: 16005, P7, SRGB Base: 16000, Weight: 0, Metric: 30`: This describes the backup path facilitated by the Classic LFA method. It utilizes the alternate next-hop address, 27.0.0.7, through the GigabitEthernet0/0/0/3 interface. The alternate next-hop router is denoted as P7, and additional details, including labels, the SRGB base, and metrics via this path, are also provided.

- `src PE5.00-00, 5.5.5.5, prefix-SID index 5, R:0 N:1 P:0 E:0 V:0 L:0, Alg:0`: This provides details about the origin router of the prefix, identified as PE5, along with information related to the prefix. This information is obtained through the IS-IS **Link State Protocol Data Units** (**LSPDUs**), as discussed in earlier chapters.

A similar output for the remaining routes and on all routers in the network would further confirm the availability of pre-computed backup paths. These paths can be utilized to reroute traffic if there is a core link failure.

> **Note**
>
> The output illustrates another path and its LFA, and the same explanation applies to it as well.
>
> This is due to the metric being set to 30 for both paths, and **Equal Cost Multi-Path** (**ECMP**) is enabled by default in the IS-IS protocol of Cisco IOS XR software.
>
> In scenarios involving ECMP, the redundant links satisfy the classic LFA inequality condition for each other. When both a classic LFA backup path and a TI-LFA backup path are available, classic LFA takes precedence.

The backup path is not only pre-computed in the control plane but also pre-installed in the data plane of the router, as evident in the following output.

Backup path in the RIB

The provided output indicates that the ECMP paths are both protected, and both serve as backup routes for each other:

```
RP/0/RP0/CPU0:P2#show route 5.5.5.5/32

Routing entry for 5.5.5.5/32
  Known via "isis IGP", distance 115, metric 30, labeled SR, type level-2
  Installed Jan  1 06:58:23.045 for 00:02:18
  Routing Descriptor Blocks
    23.0.0.3, from 5.5.5.5, via GigabitEthernet0/0/0/0, Protected, ECMP-Backup
(Local-LFA)
      Route metric is 30
    27.0.0.7, from 5.5.5.5, via GigabitEthernet0/0/0/3, Protected, ECMP-Backup
(Local-LFA)
      Route metric is 30
  No advertising protos.
RP/0/RP0/CPU0:P2#
```

The FIB derives its information from the RIB, as shown next.

Backup path in the FIB

The protection path is also installed in the FIB:

```
RP/0/RP0/CPU0:P2#show cef 5.5.5.5/32 brief
5.5.5.5/32, version 444, labeled SR, internal 0x1000001 0x8310 (ptr 0xe7ac838)
[1], 0x600 (0xda5df48), 0xa28 (0x222f94c8)
Updated Jan  1 06:58:23.081
```

```
remote adjacency to GigabitEthernet0/0/0/0
Prefix Len 32, traffic index 0, precedence n/a, priority 1
   via 23.0.0.3/32, GigabitEthernet0/0/0/0, 10 dependencies, weight 0, class
0, protected, ECMP-backup (Local-LFA) [flags 0x600]
   path-idx 0 bkup-idx 1 NHID 0x0 [0xd673730 0x0]
   next hop 23.0.0.3/32
     local label 16005      labels imposed {16005}
   via 27.0.0.7/32, GigabitEthernet0/0/0/3, 8 dependencies, weight 0, class 0,
protected, ECMP-backup (Local-LFA) [flags 0x600]
   path-idx 1 bkup-idx 0 NHID 0x0 [0xd673af0 0x0]
   next hop 27.0.0.7/32
     local label 16005      labels imposed {16005}
RP/0/RP0/CPU0:P2#
```

The same is expected in the LFIB, as shown next.

Backup path in the LFIB

The **Label Forwarding Information Base** (**LFIB**) doesn't display anything unusual; it just signifies that there's an ECMP route leading to PE5.

The LFIB can be queried using the prefix:

```
RP/0/RP0/CPU0:P2#show mpls forwarding prefix 5.5.5.5/32
Local  Outgoing    Prefix             Outgoing      Next Hop        Bytes
Label  Label       or ID              Interface                     Switched
------ ----------- ------------------ ------------  --------------- ----------
--
16005  16005       SR Pfx (idx 5)     Gi0/0/0/0     23.0.0.3        0
       16005       SR Pfx (idx 5)     Gi0/0/0/3     27.0.0.7        0
RP/0/RP0/CPU0:P2#
```

Alternatively, it can be queried using the Prefix-SID or SR-MPLS label:

```
RP/0/RP0/CPU0:P2#show mpls forwarding labels 16005
Local  Outgoing    Prefix             Outgoing      Next Hop        Bytes
Label  Label       or ID              Interface                     Switched
------ ----------- ------------------ ------------  --------------- ----------
--
16005  16005       SR Pfx (idx 5)     Gi0/0/0/0     23.0.0.3        0
       16005       SR Pfx (idx 5)     Gi0/0/0/3     27.0.0.7        0
RP/0/RP0/CPU0:P2#
```

On the P2 router, the output from the RIB, FIB, and LFIB confirms that for a route to the destination, PE5 (IP 5.5.5.5 or Node-SID 16005), an ECMP route is installed in all three tables via P3 and P7. This not only offers load balancing of traffic but also provides mutual protection if either path fails.

What if the paths are not ECMP? This is explored in the next section.

Verifying a non-ECMP backup path

This section analyzes the outputs for a non-ECMP destination, such as P8, with a loopback IP address of 8.8.8.8/32, from the P2 router.

There is a clear distinction between the primary and backup paths, as illustrated here:

```
RP/0/RP0/CPU0:P2#show route 8.8.8.8/32

Routing entry for 8.8.8.8/32
  Known via "isis IGP", distance 115, metric 20, labeled SR, type level-2
  Installed Jan  1 06:58:25.314 for 00:03:29
  Routing Descriptor Blocks
    23.0.0.3, from 8.8.8.8, via GigabitEthernet0/0/0/0, Backup (Local-LFA)
      Route metric is 30
    27.0.0.7, from 8.8.8.8, via GigabitEthernet0/0/0/3, Protected
      Route metric is 20
  No advertising protos.
RP/0/RP0/CPU0:P2#
```

The primary route from P2 to P8 is via the outbound interface, GigabitEthernet0/0/0/3, toward the next-hop 27.0.0.7, with a path cost of 20, labeled as Protected.

The backup route is via the outbound interface, GigabitEthernet0/0/0/0, through the next-hop, 23.0.0.3, with a path cost of 30, labeled as **Backup (Local-LFA)**.

Furthermore, the LFIB output for this destination shows an exclamation mark (!), indicating that it is the backup path.

> **Note**
>
> There is no exclamation mark (!) for the ECMP paths, as they all serve as primary routes and provide mutual protection.

As usual, the LFIB can be queried using the prefix, as shown here:

```
RP/0/RP0/CPU0:P2#show mpls forwarding prefix 8.8.8.8/32
Local  Outgoing    Prefix              Outgoing      Next Hop        Bytes
Label  Label       or ID               Interface                     Switched
------ ----------- ------------------- ------------ --------------- ----------
--
16008  16008       SR Pfx (idx 8)      Gi0/0/0/3    27.0.0.7        0
       16008       SR Pfx (idx
8)     Gi0/0/0/0    23.0.0.3         0             (!)
RP/0/RP0/CPU0:P2#
```

Alternatively, it can be queried using the label. In this instance, the Prefix-SID for P8 is encoded as the MPLS label:

```
RP/0/RP0/CPU0:P2#show mpls forwarding labels 16008
Local   Outgoing    Prefix               Outgoing      Next Hop          Bytes
Label   Label       or ID                Interface                       Switched
------  ----------- -------------------- ------------- ---------------- ----------
--
16008   16008       SR Pfx (idx 8)       Gi0/0/0/3     27.0.0.7          0
        16008       SR Pfx (idx
8)      Gi0/0/0/0    23.0.0.3         0                (!)
RP/0/RP0/CPU0:P2#
```

This confirms that the TI-LFA configuration has successfully established a backup path, which should be available for all routes where multiple paths exist. This is also visually apparent from *Figure 5.1*, where the P2 router has multiple links in the core; thus, if one link fails, an alternate path can take over.

The next section explains how the P2 router calculates the backup path for the destination router, P8.

Understanding backup path calculation

It's important to highlight that the outputs still specifically mention Local-LFA. This distinction arises because both classic LFA and TI-LFA are active in the topology in *Figure 5.1*. Furthermore, classic LFA inequality is met for link protection.

The following explanation details the computation and selection of the primary path and backup path:

Primary path

The primary path from P2 to P8 goes via P7, with an end-to-end path cost of 20 (P2–P7–P8), making it the shortest path.

Backup path

The following conditions are applied for the backup path.

Classic LFA

The condition for Classic LFA inequality in link protection is as follows:

```
Distance_opt(N, D) < Distance_opt(N, S) + Distance_opt(S, D)
```

Here, S = P2, D = P8, there are two neighbors, so N = P3 and P6:

- Inequality condition for P3

  S = P2, D = P8, and N = P3

  ```
  Distance_opt(P3, P8) < Distance_opt(P3, P2) + Distance_opt(P2, P8)
  20 < 10 + 20
  ```

  The condition is **satisfied**.

- Inequality condition for P6:

  S = P2, D = P8, and N = P6

  ```
  Distance_opt(P6, P8) < Distance_opt(P6, P2) + Distance_opt(P2, P8)
  20 < 10 + 20
  ```

  The condition is **satisfied**.

P2 has two neighbors that fulfill the inequality condition, but P3 is chosen because it has a lower router ID than P6. Classic LFA does not support ECMP on the backup path.

TI-LFA

The TI-LFA post-convergence path in the event of P2–P7 link failure is as follows:

- P2–P3–P7–P8, cost 30, or P2–P6–P7–P8, cost 30. If both paths have identical costs, there would be two backup paths due to ECMP.
- TI-LFA enforces the loop-free post-convergence path by encoding it as a list of segments.
- The calculation of P-Space and Q-Space is not shown for this route, as it will be covered later in the book for other scenarios.

Upon finding an alternative path through Classic LFA, it becomes the preferred choice over the TI-LFA protection method, as reflected in the output.

> Tip
> Adjust the metrics in the network topology intentionally to dissatisfy the inequality condition. Observe the subsequent transition in the protection method from Local-LFA to TI-LFA.

Summary

This chapter explores the implementation of TI-LFA within the SR-MPLS topology. It provides an overview of classic LFA, RLFA, and TI-LFA calculation methods.

Through straightforward configuration steps, TI-LFA is enabled on all IS-IS connections for every router in the network. Upon completion of the configuration process, the lab verifies the operational status of TI-LFA, ensuring its successful implementation. The verification process entails examining the output to confirm the presence of backup routes in the RIB, FIB, and LFIB for each route in a network. This verification procedure underscores the robustness of TI-LFA as a network resilience mechanism.

Furthermore, the chapter offers insights into the backup path calculation method, shedding light on the underlying principles governing the rerouting of traffic if there are link or node failures. By explaining the details of backup path calculation, the chapter equips you with a comprehensive understanding of how TI-LFA enhances network reliability and ensures uninterrupted connectivity in dynamic networking environments.

The upcoming TI-LFA labs will shed light on how segment routing SIDs encoded as labels are instrumental in calculating and installing the backup path at the PLR itself.

The next chapter will explore TI-LFA link protection FRR backup paths in more detail.

References

The following references provide additional resources and foundational specifications related to the concepts discussed in this chapter, including IP **Fast Reroute (FRR)**, **Loop-Free Alternates (LFA)**, and Segment Routing in MPLS networks:

- [RFC5286] Atlas, A., Ed. and A. Zinin, Ed., "Basic Specification for IP Fast Reroute: Loop-Free Alternates", RFC 5286, DOI 10.17487/RFC5286, September 2008, <https://www.rfc-editor.org/info/rfc5286>.

- [RFC7490] Bryant, S., Filsfils, C., Previdi, S., Shand, M., and N. So, "Remote Loop-Free Alternate (LFA) Fast Reroute (FRR)", RFC 7490, DOI 10.17487/RFC7490, April 2015, <https://www.rfc-editor.org/info/rfc7490>.

- [RFC7916] Litkowski, S., Ed., Decraene, B., Filsfils, C., Raza, K., Horneffer, M., and P. Sarkar, "Operational Management of Loop-Free Alternates", RFC 7916, DOI 10.17487/RFC7916, July 2016, <https://www.rfc-editor.org/info/rfc7916>.

- [RFC8029] Kompella, K., Swallow, G., Pignataro, C., Ed., Kumar, N., Aldrin, S., and M. Chen, "Detecting Multiprotocol Label Switched (MPLS) Data-Plane Failures", RFC 8029, DOI 10.17487/RFC8029, March 2017, <https://www.rfc-editor.org/info/rfc8029>.

- [I-D.ietf-rtgwg-segment-routing-ti-lfa] Bashandy, A., Litkowski, S., Filsfils, C., Francois, P., Decraene, B., and D. Voyer, "Topology Independent Fast Reroute using Segment Routing", Work in Progress, Internet-Draft, draft-ietf-rtgwg-segment-routing-ti-lfa-13, 16 January 2024, <https://datatracker.ietf.org/doc/html/draft-ietf-rtgwg-segment-routing-ti-lfa-13>.

- [RFC5715] Shand, M. and S. Bryant, "A Framework for Loop-Free Convergence", RFC 5715, DOI 10.17487/RFC5715, January 2010, <https://www.rfc-editor.org/info/rfc5715>.

- [RFC6571] Filsfils, C., Ed., Francois, P., Ed., Shand, M., Decraene, B., Uttaro, J., Leymann, N., and M. Horneffer, "Loop-Free Alternate (LFA) Applicability in Service Provider (SP) Networks", RFC 6571, DOI 10.17487/RFC6571, June 2012, <https://www.rfc-editor.org/info/rfc6571>.

6

Lab 5 – Zero-Segment FRR

TI-LFA FRR protection operates on the principle of creating backup paths or local repair paths on each router, for every route in the routing table. As demonstrated in the previous lab, the **Equal Cost Multi-Path** (**ECMP**) backup and non-ECMP backup paths did not have any additional labels. TI-LFA can provide 100% backup path coverage by using adjacency-SIDs. However, a backup path using an explicit list of adjacency SIDs may encounter hardware limitations, due to the maximum number of labels in the label stack. Therefore, TI-LFA backup path calculation aims to find a backup path with the fewest additional labels on top of the destination label for efficient traffic routing.

In segment routing networks, the TI-LFA mechanism aims to enforce the post-convergence backup path, with as few segments encoded as labels as possible in the label stack. This is done to reduce overhead, accelerate lookup and forwarding, and simplify operations in the hardware. Using too many labels might not be supported efficiently in the hardware.

Did you know?

The Cisco XRv9k and ASR9k platforms support a maximum of three labels for the backup path.

The upcoming labs explore various scenarios involving backup paths, ranging from zero-segment to single-segment and double-segment **Fast Reroute** (**FRR**).

Note

When the number of labels for the repair path exceeds the maximum supported value, an alternative solution will be explored later in the book.

Objective

The objective of this chapter is to explore a zero-segment FRR scenario, in which the computed backup path enables fast rerouting without requiring an additional segment encoded as a label. This is achieved through the following tasks:

1. Changing the topology to create a non-ECMP path from P2 to PE5

2. Verifying that a zero-segment backup path is computed and added to the RIB, FIB, and LFIB.

In *Figure 6.1*, the link between P2 and P3 is marked with a cross. This indicates that when the link from P2 to P3 fails, the TI-LFA mechanism should protect the traffic traversing the P2–P3 link by quickly rerouting it to an alternative path.

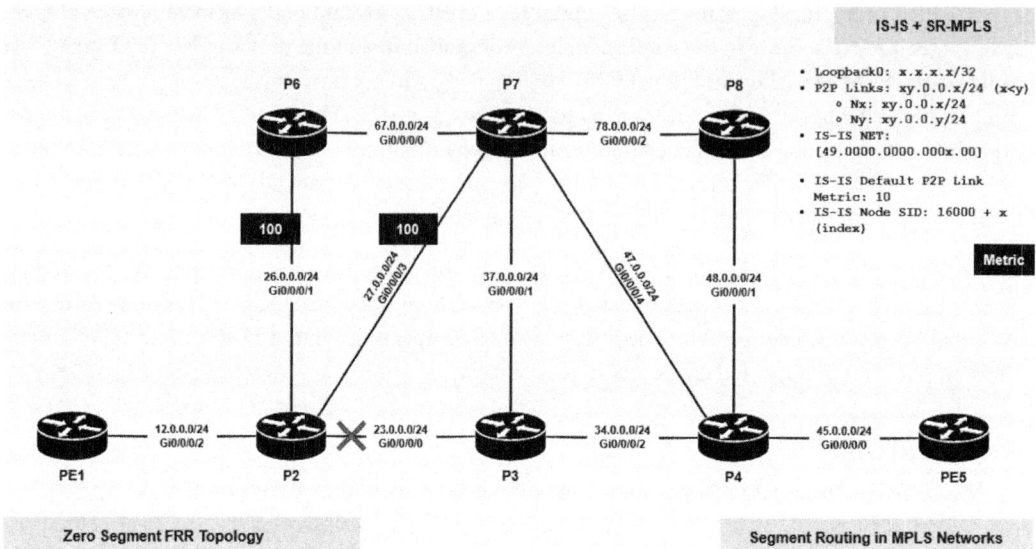

Figure 6.1 – Zero-segment FRR

To facilitate this, TI-LFA on P2 already calculates and installs the backup path for such potential failures in the network.

Next, we will adjust the configuration on the routers to reflect the topology described in *Figure 6.1*.

Configuration

Modify the metric in the following links to create a non-ECMP path from P2 to PE:

- On the P2 router, increase the metric from 10 to 100 on the GigabitEthernet0/0/0/1 and GigabitEthernet0/0/0/3 interfaces toward the routers P6 and P7, respectively:

P2

```
router isis IGP
interface GigabitEthernet0/0/0/1
  address-family ipv4 unicast
   metric 100
  !
 !
interface GigabitEthernet0/0/0/3
  address-family ipv4 unicast
   metric 100
```

- On the P6 router, increase the metric from 10 to 100 on the GigabitEthernet0/0/0/1 interface toward P2:

P6

```
router isis IGP
interface GigabitEthernet0/0/0/1
  address-family ipv4 unicast
   metric 100
```

- On the P7 router, increase the metric from 10 to 100 on the GigabitEthernet0/0/0/3 interface toward P2:

P7

```
router isis IGP
interface GigabitEthernet0/0/0/3
  address-family ipv4 unicast
   metric 100
```

The increase in the metrics results in a single best path from P2 to PE5, which is through P3.

The next section verifies the availability of the backup path on P2 if the P2–P3 link fails.

Verification

The following section verifies that the zero-segment FRR path is installed in the RIB, LFIB, and FIB.

Verifying max label support

The displayed output indicates that the XRv9K platform supports a maximum of three labels, and this holds true for the ASR9k platform as well:

```
RP/0/RP0/CPU0:P2#show isis interface Gi0/0/0/0 | i MPLS
    MPLS Max Label Stack:    3/3/10/10 (PRI/BKP/SRTE/SRAT)
    MPLS LDP Sync (L1/L2):  Disabled/Disabled
RP/0/RP0/CPU0:P2#
```

This implies that, in addition to the destination node label, the backup path can accommodate a maximum of two additional labels.

Verifying the zero-segment backup path

The following backup path output indicates the absence of any additional labels.

The primary path goes through P3, using the Gi0/0/0/0 interface. In the event of a failure on this link, the backup path reroutes through P7.

This is evident in the RIB, FIB, and LFIB.

Backup path in RIB

The backup path computation in the following output shows the Prefix-SID destination router, also known as Node-SID, as the only label:

```
RP/0/RP0/CPU0:P2#show isis fast-reroute detail 5.5.5.5/32

L2 5.5.5.5/32 [30/115] Label: 16005, medium priority
    Installed Jan 01 07:04:59.930 for 00:00:38
      via 23.0.0.3, GigabitEthernet0/0/0/0, Label: 16005, P3, SRGB Base: 16000,
Weight: 0
       Backup path: LFA, via 27.0.0.7, GigabitEthernet0/0/0/3, Label: 16005,
P7, SRGB Base: 16000, Weight: 0, Metric: 120
       P: No, TM: 120, LC: No, NP: Yes, D: Yes, SRLG: Yes
     src PE5.00-00, 5.5.5.5, prefix-SID index 5, R:0 N:1 P:0 E:0 V:0 L:0,
Alg:0
RP/0/RP0/CPU0:P2#
```

The TI-LFA backup path, as computed and shown in the preceding output, is also injected into the routing table, as detailed here:

```
RP/0/RP0/CPU0:P2#show route 5.5.5.5/32

Routing entry for 5.5.5.5/32
  Known via "isis IGP", distance 115, metric 30, labeled SR, type level-2
```

```
    Installed Jan  1 07:04:59.930 for 00:00:53
    Routing Descriptor Blocks
      23.0.0.3, from 5.5.5.5, via GigabitEthernet0/0/0/0, Protected
        Route metric is 30
      27.0.0.7, from 5.5.5.5, via GigabitEthernet0/0/0/3, Backup (Local-LFA)
        Route metric is 120
    No advertising protos.
RP/0/RP0/CPU0:P2#
```

The P7 router can send traffic to the destination node, PE5, through its downstream path without the need for additional labels. In this case, only the destination label, 16005, is required for P7 to make the correct forwarding decision.

Backup path in FIB

The only label imposed on the backup path is the destination router, Node-SID:

```
RP/0/RP0/CPU0:P2#show cef 5.5.5.5/32 brief
5.5.5.5/32, version 522, labeled SR, internal 0x1000001 0x8310 (ptr 0xe7ac838)
[1], 0x600 (0xda5df48), 0xa28 (0x222f98e8)
Updated Jan  1 07:04:59.939
remote adjacency to GigabitEthernet0/0/0/0
Prefix Len 32, traffic index 0, precedence n/a, priority 1
   via 23.0.0.3/32, GigabitEthernet0/0/0/0, 14 dependencies, weight 0, class
0, protected [flags 0x400]
    path-idx 0 bkup-idx 1 NHID 0x0 [0xd673730 0x0]
    next hop 23.0.0.3/32
     local label 16005     labels imposed {16005}
   via 27.0.0.7/32, GigabitEthernet0/0/0/3, 12 dependencies, weight 0, class
0, backup (Local-LFA) [flags 0x300]
    path-idx 1 NHID 0x0 [0x220323d0 0x0]
    next hop 27.0.0.7/32
    remote adjacency
     local label 16005     labels imposed {16005}
RP/0/RP0/CPU0:P2#
```

The preceding FIB output and the detailed LFIB output indicate that there are no extra labels in the label stack on the backup path. This implies that on both the primary and backup paths, the traffic would be loop-free without any additional forwarding information.

Backup path in LFIB

The LFIB output shows only the destination label in the label stack, with no additional segments:

```
RP/0/RP0/CPU0:P2#show mpls forwarding prefix 5.5.5.5/32 detail
Local  Outgoing    Prefix              Outgoing      Next Hop         Bytes
Label  Label       or ID               Interface                      Switched
------ ----------- ------------------- ------------- ---------------- ----------
--
16005  16005       SR Pfx (idx 5)      Gi0/0/0/0     23.0.0.3         0
       Updated: Jan  1 07:04:59.939
       Path Flags: 0x400 [ BKUP-IDX:1 (0xd673730) ]
       Version: 522, Priority: 1
       Label Stack (Top -> Bottom): { 16005 }
       NHID: 0x0, Encap-ID: N/A, Path idx: 0, Backup path idx: 1, Weight: 0
       MAC/Encaps: 4/8, MTU: 1500
       Outgoing Interface: GigabitEthernet0/0/0/0 (ifhandle 0x01000020)
       Packets Switched: 0

       16005       SR Pfx (idx
5)     Gi0/0/0/3   27.0.0.7            0              (!)
       Updated: Jan  1 07:04:59.939
       Path Flags: 0x300 [ IDX:1 BKUP, NoFwd ]
       Version: 522, Priority: 1
       Label Stack (Top -> Bottom): { 16005 }
       NHID: 0x0, Encap-ID: N/A, Path idx: 1, Backup path idx: 0, Weight: 0
       MAC/Encaps: 4/8, MTU: 1500
       Outgoing Interface: GigabitEthernet0/0/0/3 (ifhandle 0x01000048)
       Packets Switched: 0
       (!): FRR pure backup

   Traffic-Matrix Packets/Bytes Switched: 0/0
RP/0/RP0/CPU0:P2#
```

Next, it is important to understand how this zero-segment backup path was calculated and finalized.

Understanding backup path calculation

The following topology diagram shows the backup path.

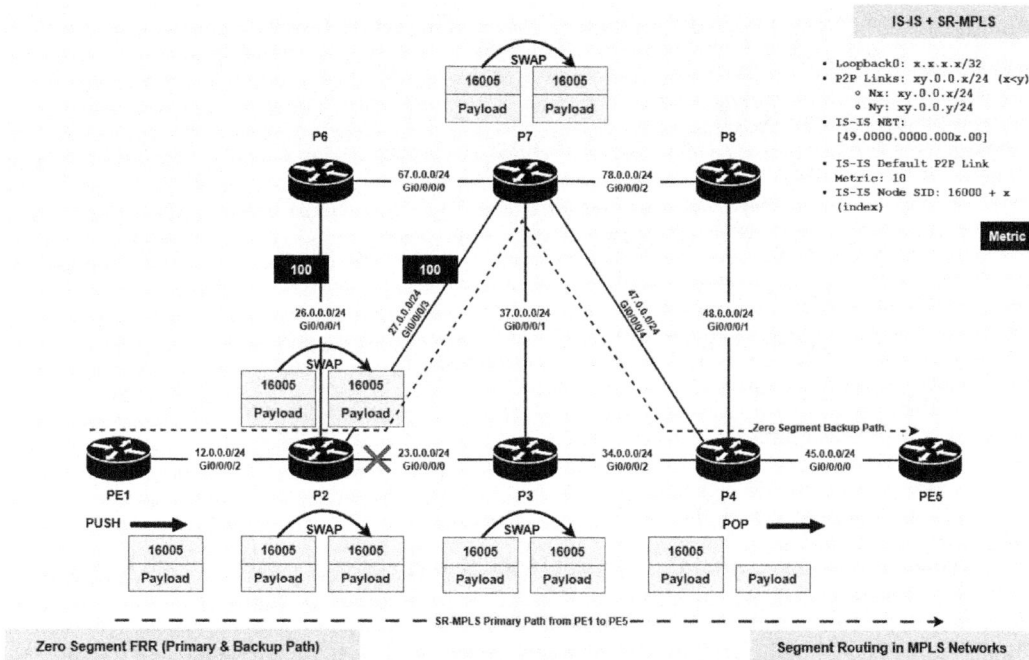

Figure 6.2 – Zero-Segment FRR (primary and backup path)

The following explanation details the computation and selection of the primary path and backup path.

Primary path

The primary path from P2 to PE5 goes via P3, with the end-to-end path costing 30 (P3–P4–PE5), making it the shortest path.

Backup path

Moreover, the backup path is classified as Local-LFA because the conditions for link protection inequality, explained in the previous chapter, remain applicable in this context.

Classic LFA

As per RFC 5286, Inequality 1: Loop-Free Criterion is defined as, "*A neighbor N can provide a* **loop-free alternate (LFA)** *if, and only if,* Distance_opt(N, D) < Distance_opt(N, S) + Distance_opt(S, D)"

Here, S is used to indicate the calculating router. N is a neighbor of S, and N is used as an abbreviation when only one neighbor is being discussed; otherwise, N_i is used to represent more than one neighbor. D is the destination under consideration.

In this case, S = P2 and D = PE5, as there are two neighbors, so N = P6 and P7. The inequality conditions for P6 and P7 are as follows:

- **Inequality condition for P6**:

  S = P2, D = PE5, and N = P6

  ```
  Distance_opt(P6, PE5) < Distance_opt(P6, P2) + Distance_opt(P2, PE5)
  30 (P7 - P4 - PE5) < 100 + 30 (P3 - P4 - PE5)
  ```

  The condition is **satisfied**

- **Inequality condition for P7**:

  S = P2, D = PE5, and N = P7

  ```
  Distance_opt(P7, PE5) < Distance_opt(P7, P2) + Distance_opt(P2, PE5)
  20 (P4 - PE5) < 100 + 30 (P3 - P4 - PE5)
  ```

  The condition is **satisfied**

P2 has two neighbors that meet the inequality condition; however, the path through P7 is shorter than the one through P6, which is why it is selected as the preferred backup path.

In the event of a P2–P3 link failure, the P2 router redirects traffic to the pre-computed backup path without requiring any additional label. This is because the reachability from P7 to PE5 does not involve the P2–P3 link. Therefore, the outgoing label on the backup path is also the label for the destination node, PE5.

An MPLS path is a stack of one or more labels. While IP routing and forwarding always follow the RIB, MPLS-labeled forwarding follows the LFIB. It is possible to trace the backup path, as shown next, to ensure that it is ready to carry the fast-rerouted if there is a failure.

Verifying the backup path traceroute

In a regular traceroute, the primary path is used. The MPLS traceroute and ping OAM tool have an extension that incorporates the use of nil-fec. This feature enables you to test with any label stack.

This method serves as a way to test the backup path, which only comes into play in case of a failure.

The essential elements for conducting the nil-fec test on the P2 router for the destination router, PE5, can be extracted from the aforementioned RIB, FIB, or LFIB outputs.

The input required for the nil-fec traceroute comprises three essential components:

- **Label stack (top to bottom)**: There is only the destination label 16005 in the label stack of the backup path.

- **Outgoing interface**: The outgoing interface on router P2 for the backup path to PE5 is `GigabitEthernet0/0/0/3`, which points toward router P7.
- **Outgoing interface next-hop IP address**: The next-hop IP address on the backup path from the P2 router toward the P7 router, on the `GigabitEthernet0/0/0/3` interface, is `27.0.0.7`.

All of the preceding components are combined in a `traceroute` command, as shown below, to validate the labeled forwarding through the network from the local router, P2, to the destination router, PE5:

```
RP/0/RP0/CPU0:P2#traceroute sr-mpls nil-fec labels 16005 output interface
GigabitEthernet0/0/0/3 nexthop 27.0.0.7 verbose

Tracing MPLS Label Switched Path with Nil FEC with labels [16005], timeout is
2 seconds

Codes: '!' - success, 'Q' - request not sent, '.' - timeout,
  'L' - labeled output interface, 'B' - unlabeled output interface,
  'D' - DS Map mismatch, 'F' - no FEC mapping, 'f' - FEC mismatch,
  'M' - malformed request, 'm' - unsupported tlvs, 'N' - no rx label,
  'P' - no rx intf label prot, 'p' - premature termination of LSP,
  'R' - transit router, 'I' - unknown upstream index,
  'X' - unknown return code, 'x' - return code 0

Type escape sequence to abort.

  0 27.0.0.2 27.0.0.7 MRU 1500 [Labels: 16005/explicit-null Exp: 0/0]
L 1 27.0.0.7 47.0.0.4 MRU 1500 [Labels: 16005/explicit-null Exp: 0/0] 17 ms,
ret code 8
L 2 47.0.0.4 45.0.0.5 MRU 1500 [Labels: implicit-null/explicit-null Exp: 0/0]
11 ms, ret code 8
! 3 45.0.0.5 27 ms, ret code 3
RP/0/RP0/CPU0:P2#
```

The successful outcome of the traceroute confirms that the backup path is not only calculated and installed in the control plane but also in the forwarding plane. If there is a failure, the traffic can be fast rerouted through this path.

Before moving on to the next lab, let's restore the topology, as shown in the next section.

Restoring the topology

It is important to normalize the metric on the P2–P6 and P2–P7 links back from 100 to 10 so that the topology in *Figure 6.1* can be prepared for the next chapter. This is shown as follows:

- On the router P2, increase the metric from 100 to 10 on the GigabitEthernet0/0/0/1 and GigabitEthernet0/0/0/3 interfaces toward the P6 and P7 routers, respectively:

P2

```
router isis IGP
interface GigabitEthernet0/0/0/1
  address-family ipv4 unicast
   metric 10
   !
 !
interface GigabitEthernet0/0/0/3
  address-family ipv4 unicast
   metric 10
```

- On the P6 router, increase the metric from 100 to 10 on the GigabitEthernet0/0/0/1 interface toward P2:

P6

```
router isis IGP
interface GigabitEthernet0/0/0/1
  address-family ipv4 unicast
   metric 10
```

- On the P7 router, increase the metric from 100 to 10 on the GigabitEthernet0/0/0/3 interface toward P2:

P7

```
router isis IGP
interface GigabitEthernet0/0/0/3
  address-family ipv4 unicast
   metric 10
```

The zero-segment fast reroute scenario is concluded.

Summary

In this chapter, the focus was on zero-segment FRR, a scenario where MPLS backup paths are established without requiring any additional segment or label in the label stack. The primary objective was to ensure that if there is a link or node failure, a swift and efficient rerouting mechanism is in place to maintain network connectivity.

To achieve this, the topology was modified to create a non-ECMP path from P2 to PE5. This ensured that there was a distinct path available for fast rerouting purposes. The lab then proceeded to verify that the zero-segment backup path was correctly computed and added to the **Routing Information Base (RIB)**, **Forwarding Information Base (FIB)**, and **Label Forwarding Information Base (LFIB)**.

The emphasis was placed on understanding the backup path calculation process and the rationale behind choosing one method over another.

Once the lab objectives were achieved and the functionality of the backup path was verified, the network topology was reverted to its original configuration. This ensured that the network was ready for subsequent labs, marking the conclusion of the zero-segment FRR scenario.

The next chapter will explore the single-segment FFR scenario.

References

The following references provide foundational specifications and further readings on IP Fast Reroute, Segment Routing, and related MPLS technologies discussed in this chapter.

- *[RFC5286] Atlas, A., Ed. and A. Zinin, Ed., "Basic Specification for IP Fast Reroute: Loop-Free Alternates", RFC 5286, DOI 10.17487/RFC5286, September 2008, <https://www.rfc-editor.org/info/rfc5286>.*

- *[RFC7490] Bryant, S., Filsfils, C., Previdi, S., Shand, M., and N. So, "Remote Loop-Free Alternate (LFA) Fast Reroute (FRR)", RFC 7490, DOI 10.17487/RFC7490, April 2015, <https://www.rfc-editor.org/info/rfc7490>.*

- *[RFC7916] Litkowski, S., Ed., Decraene, B., Filsfils, C., Raza, K., Horneffer, M., and P. Sarkar, "Operational Management of Loop-Free Alternates", RFC 7916, DOI 10.17487/RFC7916, July 2016, <https://www.rfc-editor.org/info/rfc7916>.*

- *[RFC8029] Kompella, K., Swallow, G., Pignataro, C., Ed., Kumar, N., Aldrin, S., and M. Chen, "Detecting Multiprotocol Label Switched (MPLS) Data-Plane Failures", RFC 8029, DOI 10.17487/RFC8029, March 2017, <https://www.rfc-editor.org/info/rfc8029>.*

- *[I-D.ietf-rtgwg-segment-routing-ti-lfa] Bashandy, A., Litkowski, S., Filsfils, C., Francois, P., Decraene, B., and D. Voyer, "Topology Independent Fast Reroute using Segment Routing", Work in Progress, Internet-Draft, draft-ietf-rtgwg-segment-routing-ti-lfa-13, 16 January 2024, <https://datatracker.ietf.org/doc/html/draft-ietf-rtgwg-segment-routing-ti-lfa-13>.*

- *[RFC5715] Shand, M. and S. Bryant, "A Framework for Loop-Free Convergence", RFC 5715, DOI 10.17487/RFC5715, January 2010, <https://www.rfc-editor.org/info/rfc5715>.*

- *[RFC6571] Filsfils, C., Ed., Francois, P., Ed., Shand, M., Decraene, B., Uttaro, J., Leymann, N., and M. Horneffer, "Loop-Free Alternate (LFA) Applicability in Service Provider (SP) Networks", RFC 6571, DOI 10.17487/RFC6571, June 2012, <https://www.rfc-editor.org/info/rfc6571>.*

7

Lab 6 – Single-Segment FRR

Building on the previous lab, this lab explores a scenario where a loop-free backup path cannot be guaranteed without an additional segment encoded as a label in the label stack. However, the **Topology Independent Loop-Free Alternate** (**TI-LFA**) post-convergence path checks whether a loop-free alternative route to the destination can be achieved, by using one additional segment or label, hence called the single-segment **fast reroute** (**FRR**) scenario.

Objective

The objective of this lab is to investigate a single-segment FRR scenario where TI-LFA provides a backup path through one additional segment encoded as a label in the backup path label stack.

This is accomplished by the tasks outlined here:

1. The topology is modified by shutting down the links on router P7, as illustrated in *Figure 7.1*
2. Verify whether a single-segment TI-LFA backup path is provided and installed in the **Routing Information Base** (**RIB**), **Forwarding Information Base** (**FIB**), and **Label Forwarding Information Base** (**LFIB**)

After shutting down the necessary links on router P7, the network topology for the purpose of this chapter would appear as described in *Figure 7.1*.

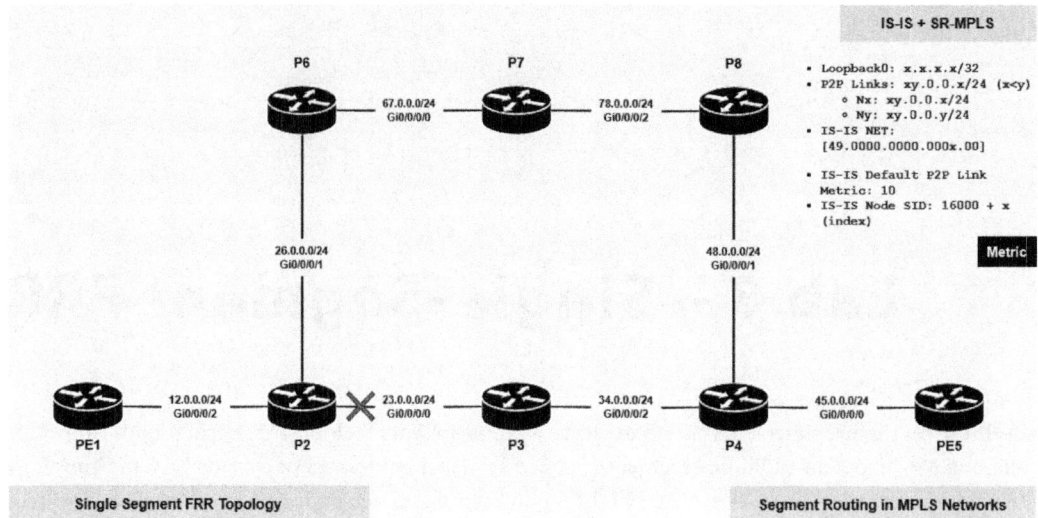

Figure 7.1 – Single-segment FRR

The first step is to apply the necessary configuration to achieve the preceding state, as shown next.

Configuration

To create the ring topology in *Figure 7.1*, the links toward P2, P3, and P4 must be shut down on router P7.

The interfaces can be shut down under the IS-IS context, as shown in the following. This action would trigger an IS-IS update from router P7, withdrawing the link state information for those links and excluding them from the active topology.

P7

```
router isis IGP
interface GigabitEthernet0/0/0/1
  shutdown
!
interface GigabitEthernet0/0/0/3
  shutdown
!
interface GigabitEthernet0/0/0/4
  shutdown
```

As visually evident from *Figure 7.1*, after shutting down the specified links, the primary path from P2 to PE5 is via P3 due to the lower end-to-end path cost. The next step is to verify whether TI-LFA was able to offer a backup path, as described in the following section.

Verification

If the link between P2 and P3 goes down or router P3 becomes unavailable due to a node failure, the only remaining path for P2 to reach PE5 is through the next hop, P6. This path represents the TI-LFA post-convergence route in the next section.

Verify single-segment backup path

The following RIB output indicates that the backup path now uses `TI-LFA (link)`, signifying the default link protection method, along with an additional label, `16008`. This label corresponds to the Node-SID of P8.

Backup path in RIB

The backup path is on the post-convergence path if the P2-P3 link fails, as shown here:

```
RP/0/RP0/CPU0:P2#show isis fast-reroute detail 5.5.5.5/32


L2 5.5.5.5/32 [30/115] Label: 16005, medium priority
    Installed Jan 01 07:14:44.417 for 00:00:13
      via 23.0.0.3, GigabitEthernet0/0/0/0, Label: 16005, P3, SRGB Base: 16000,
Weight: 0
          Backup path: TI-LFA (link), via 26.0.0.6, GigabitEthernet0/0/0/1 P6,
SRGB Base: 16000, Weight: 0, Metric: 50
            P node: P8.00 [8.8.8.8], Label: 16008
            Prefix label: 16005
            Backup-src: PE5.00
          P: No, TM: 50, LC: No, NP: No, D: No, SRLG: Yes
      src PE5.00-00, 5.5.5.5, prefix-SID index 5, R:0 N:1 P:0 E:0 V:0 L:0,
Alg:0
RP/0/RP0/CPU0:P2#
```

The routing table output shows the presence of a repair node, `8.8.8.8`, along the backup path, corresponding to the P node from the extended P-Space of P2, identified in the previous output:

```
RP/0/RP0/CPU0:P2#show route 5.5.5.5/32


Routing entry for 5.5.5.5/32
  Known via "isis IGP", distance 115, metric 30, labeled SR, type level-2
  Installed Jan  1 07:14:44.418 for 00:00:27
  Routing Descriptor Blocks
    23.0.0.3, from 5.5.5.5, via GigabitEthernet0/0/0/0, Protected
      Route metric is 30
    26.0.0.6, from 5.5.5.5, via GigabitEthernet0/0/0/1, Backup (TI-LFA)
      Repair Node(s): 8.8.8.8
      Route metric is 50
```

```
    No advertising protos.
RP/0/RP0/CPU0:P2#
```

In the Zero-Segment FRR path of the previous lab, the backup path didn't require any extra information for loop-free reachability to the destination. However, in the single-segment FRR scenario, a repair path tunnel is established to the repair node on the backup path to ensure loop-free reachability to the destination.

In the process of reaching the repair node, an additional label is appended to the label stack. The same is downloaded into the FIB, next.

Backup path in FIB

The FIB output confirms the details checked earlier:

```
RP/0/RP0/CPU0:P2#show cef 5.5.5.5/32 brief
5.5.5.5/32, version 624, labeled SR, internal 0x1000001 0x8310 (ptr 0xe7ac838)
[1], 0x600 (0xda5df48), 0xa28 (0x222f9208)
Updated Jan  1 07:14:44.426
remote adjacency to GigabitEthernet0/0/0/0
Prefix Len 32, traffic index 0, precedence n/a, priority 1
   via 23.0.0.3/32, GigabitEthernet0/0/0/0, 14 dependencies, weight 0, class
0, protected [flags 0x400]
    path-idx 0 bkup-idx 1 NHID 0x0 [0xd673dc0 0x0]
    next hop 23.0.0.3/32
     local label 16005        labels imposed {16005}
   via 26.0.0.6/32, GigabitEthernet0/0/0/1, 11 dependencies, weight 0, class
0, backup (TI-LFA) [flags 0xb00]
    path-idx 1 NHID 0x0 [0x22032470 0x0]
    next hop 26.0.0.6/32, Repair Node(s): 8.8.8.8
    remote adjacency
     local label 16005        labels imposed {16008 16005}
RP/0/RP0/CPU0:P2#
```

The LFIB corresponds to the same, as shown next.

Backup path in LFIB

The RIB output indicated that router P8 is on the backup path via router P6. Traffic needs to be tunneled to P8 first, as its Node-SID label was also shown. The subsequent LFIB and FIB output verify that the label stack has been appended accordingly to reflect the information calculated in the RIB by TI-LFA:

```
RP/0/RP0/CPU0:P2#show mpls forwarding prefix 5.5.5.5/32 detail
Local  Outgoing    Prefix             Outgoing      Next Hop        Bytes
Label  Label       or ID              Interface                     Switched
------ ----------- ------------------ ------------- --------------- -----------
16005  16005       SR Pfx (idx 5)     Gi0/0/0/0     23.0.0.3        0
```

```
       Updated: Jan  1 07:14:44.428
       Path Flags: 0x400 [  BKUP-IDX:1 (0xd673dc0) ]
       Version: 624, Priority: 1
       Label Stack (Top -> Bottom): { 16005 }
       NHID: 0x0, Encap-ID: N/A, Path idx: 0, Backup path idx: 1, Weight: 0
       MAC/Encaps: 4/8, MTU: 1500
       Outgoing Interface: GigabitEthernet0/0/0/0 (ifhandle 0x01000020)
       Packets Switched: 0


         16008        SR Pfx (idx
  5)     Gi0/0/0/1    26.0.0.6          0                  (!)
       Updated: Jan  1 07:14:44.428
       Path Flags: 0xb00 [  IDX:1 BKUP, NoFwd ]
       Version: 624, Priority: 1
       Label Stack (Top -> Bottom): { 16008 16005 }
       NHID: 0x0, Encap-ID: N/A, Path idx: 1, Backup path idx: 0, Weight: 0
       MAC/Encaps: 4/12, MTU: 1500
       Outgoing Interface: GigabitEthernet0/0/0/1 (ifhandle 0x01000058)
       Packets Switched: 0
       (!): FRR pure backup

   Traffic-Matrix Packets/Bytes Switched: 0/0
 RP/0/RP0/CPU0:P2#
```

The backup path is now composed of a stack of labels: 16008 and 16005. This indicates that, on this path, the traffic coming from P2 would first be directed to P8, which would then forward it to the destination node, PE5.

The outputs confirm that, on router P2, the TI-LFA backup path or local repair path to the destination PE5, in the event of a failure on the P2-P3 link, traverses via P6, P7, and P8. This path ensures loop-free operation: traffic is initially forwarded to router P8, with the Node-SID of P8 encoded as the top label. Upon arrival at P8, the second label to PE5 becomes the top label, guiding the traffic directly to its destination.

In the context of a lab environment with a small number of routers, the steps involved appear straightforward and manageable. However, in production networks that span hundreds of nodes and thousands of links, the complexity increases significantly. Visual deductions and path calculations become more intricate due to the scale and interdependencies among network elements.

The next section will delve into the TI-LFA post-convergence path calculation, illustrating how this process ensures the availability of reliable backup paths in such expansive network environments.

Understanding backup path calculation

In the previous chapter, the Classic-LFA backup path was chosen whenever available, prioritizing it over the TI-LFA backup path. This decision was due to the fulfillment of the inequality condition, ensuring a straightforward backup path calculation.

In the current lab, however, the Classic-LFA inequality condition was not met. Consequently, a TI-LFA backup path was sought, and the following section provides a detailed explanation of this method.

The following topology diagram is relevant to the TI-LFA calculation explanation that follows.

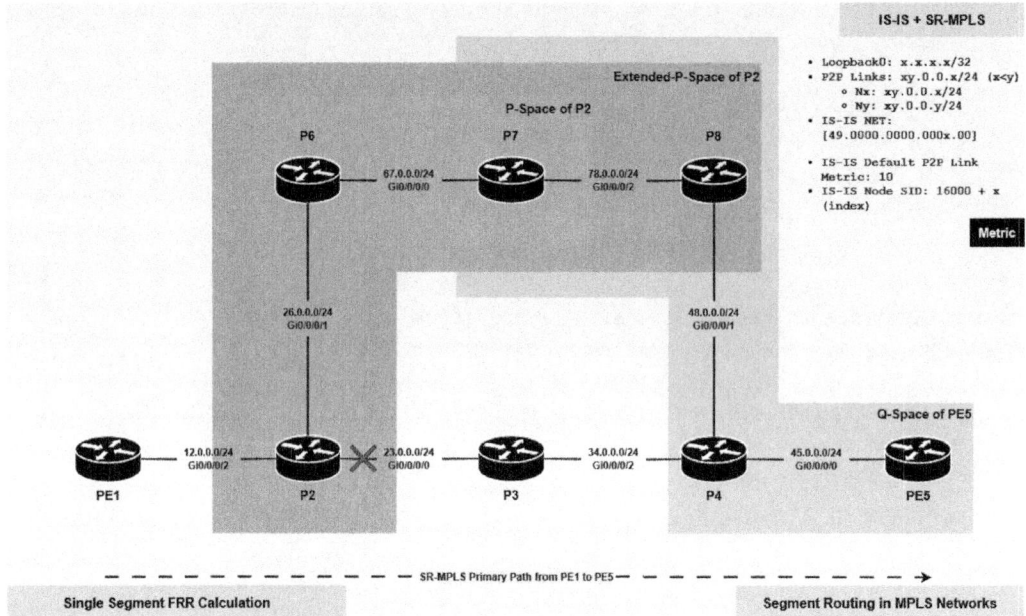

Figure 7.2 – Single-segment FRR path calculation

The following explanation details the computation and selection of the primary path and backup path.

Primary path

The primary path from P2 to PE5 goes via P3, with the end-to-end path, `cost of 30 (P2 - P3 - P4 - PE5)`, making it the shortest path.

Backup path

Moreover, the backup path is classified as `TI-LFA (link)` because the classic LFA for link protection inequality condition is not met here.

Classic LFA

As per RFC 5286, Inequality 1, the loop-free criterion *"A neighbor N can provide a loop-free alternate (LFA) if and only if* `Distance_opt(N, D) < Distance_opt(N, S) + Distance_opt(S, D)`*".*

Here, S is used to indicate the calculating router. N is a neighbor of S; N is used as an abbreviation when only one neighbor is being discussed, else N_i for more neighbors. D is the destination under consideration.

In this case, S = P2, D = PE5, and N = P6.

Inequality condition for P6:

```
S = P2, D = PE5 and N = P6
Distance_opt(P6, PE5) < Distance_opt(P6, P2) + Distance_opt(P2, PE5)
40 (P7 - P8 - P4 - PE5) < (10 + 30) (P3 - P4 - PE5)
```

The condition is NOT satisfied. Therefore, the Classic-LFA backup path is unavailable.

TI-LFA

TI-LFA establishes a backup path by calculating a post-convergence route and identifying P and Q nodes along it, as mentioned in *Chapter 5*. The repair path is then guided toward the PQ or P then the Q node, ensuring a path to the destination. This approach guarantees 100% coverage in all situations.

In this scenario, the post-convergence path would be P2 - P6 - P7 - P8 - P4 - PE5 with a cost of 50.

In addressing the link failure on P2-P3, TI-LFA accomplishes this by collecting the following information:

- **P-Space of P2 along the post-convergence path**: The group of routers on the post-convergence path reachable from P2 through the shortest path tree, excluding the traversal of P2-P3, forms the P-Space of P2. This is determined by calculating the shortest path tree with P2 as the root and removing the subtree accessed via the P2-P3 link (including the ECMP nodes, such as P8).

  In *Figure 7.2*, these routers are P6 and P7.

  The pre-convergence path cost from P2 to P6 is 10, and from P2 to P7 via P6 is 20. The path from P2 to both routers does not traverse the P2–P3 link even before failure.

- **Extended P-Space of P2 along the post-convergence path**: The extended P-Space of P2 is the combination of P2's P-Space and the P-Space of P2's neighbors along the post-convergence path. This can be determined by calculating the shortest path tree rooted at each of P2's neighbors and removing the subtree accessed via the P2-P3 link (including the ECMP nodes).

  There is only one neighbor of P2 on the post-convergence path, which is P6.

  The P-Space of P6 comprises P7 and P8. Regardless of the failure of the P2-P3 link, router P6 can reach router P7 with a cost of 10 and router P8 via P7 with a cost of 20.

  Therefore, the extended P-Space of P2 along the post-convergence path includes P6, P7, and P8.

- **Q-Space of PE5 along the post-convergence path**: The group of routers along the post-convergence path, through which router PE5 can be reached via the shortest path tree without traversing the P2-P3 link, forms the Q-Space of PE5. This is achieved by computing a reverse shortest path tree rooted at PE5 and removing the subtree reached via the P2-P3 links (including the ECMP nodes).

  According to the topology depicted in *Figure 7.2*, routers P7, P8, and P4 can reach router PE5 without traversing the P2-P3 links. The path cost from P7 to PE5 via P8 and P4 is 30, from P8 to PE5 via P4 is 20, and from P4 to PE5 is 10.

  The Q-Space of PE5 comprises P7, P8, and P4.

  The TI-LFA repair path is a sequence of segments (a stack of labels) along the post-convergence path on the outgoing backup interface. This path is determined by the intersection of the extended P-Space of P2 and the Q-Space of PE5.

  The nodes that overlap between the extended P-Space and Q-Space are designated as PQ nodes, P7 and P8, in this scenario.

  The endpoint for the repair tunnel is chosen from the set of PQ nodes. In the Cisco IOS-XR implementation of TI-LFA, the selection is based on the deeper downstream node in the post-convergence path, and in this case, it is P8.

  This is evident in the output where the outgoing label on the backup path from the **Point of Local Repair (PLR)** P2 to the destination PE5 is the Node-SID of P8. Upon examining the label stack, it becomes apparent that the Node-SID of the destination PE5 is at the bottom of the stack.

  In the event of a P2-P3 link failure, router P2 will redirect the traffic to the backup path, destined for P8. The path from P2 to P8 does not traverse the P2-P3 link. Upon receiving the traffic, P8 observes an additional label and consults its LFIB to forward it. The subsequent path from P8 to PE5 also avoids the P2-P3 link. Thus, the TI-LFA backup path ensures loop-free forwarding.

Using the preceding method, TI-LFA ensures 100% coverage by establishing a backup path for each route in the routing table of all routers.

The backup path traceroute presented in the next section validates that the TI-LFA calculation is loop-free in the control plane and avoids traversal of the P2-P3 link.

Verifying backup path traceroute

The essential components necessary for conducting the `nil-fec` traceroute are derived from the previously mentioned RIB, FIB, or LFIB outputs and are as follows:

- Label stack (top to bottom): `16008`, `16005`

- Outgoing interface: `GigabitEthernet0/0/0/1` toward P6

- Next-hop IP address of outgoing interface: `26.0.0.6` to P6

Using these components together in the traceroute `nil-fec` command on router P2, as shown next, validates the availability of the single-segment TI-LFA backup path in the network:

```
RP/0/RP0/CPU0:P2#traceroute sr-mpls nil-fec labels 16008,16005 output
interface GigabitEthernet0/0/0/1 nexthop 26.0.0.6

Tracing MPLS Label Switched Path with Nil FEC with labels [16008,16005],
timeout is 2 seconds

Codes: '!' - success, 'Q' - request not sent, '.' - timeout,
  'L' - labeled output interface, 'B' - unlabeled output interface,
  'D' - DS Map mismatch, 'F' - no FEC mapping, 'f' - FEC mismatch,
  'M' - malformed request, 'm' - unsupported tlvs, 'N' - no rx label,
  'P' - no rx intf label prot, 'p' - premature termination of LSP,
  'R' - transit router, 'I' - unknown upstream index,
  'X' - unknown return code, 'x' - return code 0

Type escape sequence to abort.

  0 26.0.0.2 MRU 1500 [Labels: 16008/16005/explicit-null Exp: 0/0/0]
L 1 26.0.0.6 MRU 1500 [Labels: 16008/16005/explicit-null Exp: 0/0/0] 42 ms
L 2 67.0.0.7 MRU 1500 [Labels: implicit-null/16005/explicit-null Exp: 0/0/0]
10 ms
L 3 78.0.0.8 MRU 1500 [Labels: 16005/explicit-null Exp: 0/0] 38 ms
L 4 48.0.0.4 MRU 1500 [Labels: implicit-null/explicit-null Exp: 0/0] 20 ms
! 5 45.0.0.5 19 ms
RP/0/RP0/CPU0:P2#
```

The following topology diagram shows the primary and the backup paths:

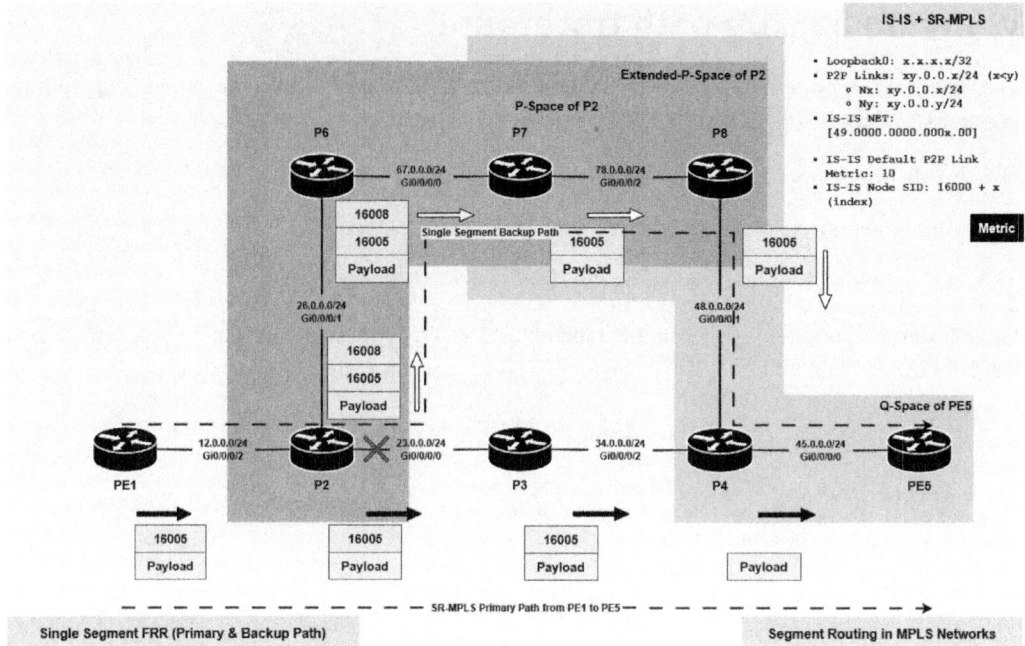

Figure 7.3 – Single-segment FRR (primary and backup paths)

The following trace output illustrates the traversal of the MPLS LSP from router P2 to destination PE5, as per the path calculated by TI-LFA and as shown in *Figure 7.3*:

- **Hop 0**: The traceroute originates from router P2 with an MPLS label stack of 16008/16005/ explicit-null. The inclusion of explicit-null ensures that the penultimate router does not remove the MPLS labels prematurely.

- **Hop 1**: The trace exits router P6, which inspects the top MPLS label, consults its LFIB, swaps the top label with the outgoing label (16008), and forwards the packet based on the LFIB's next-hop and outgoing interface information.

- **Hop 2**: The trace proceeds through router P7, which examines the top MPLS label, checks its LFIB, and determines that it is the penultimate router for P8. It pops the top label (hence, implicit-null) and forwards the packet along the MPLS label-switched path.

- **Hop 3**: The trace moves through router P8, receiving the top MPLS label as 16005. P8 consults its LFIB, swaps the outgoing label to 16005, and forwards the packet along the MPLS label-switched path.

- **Hop 4**: The trace passes through router P4, which serves as the penultimate router for PE5. The MPLS label stack includes implicit-null and explicit-null. This configuration ensures that P4 forwards the packet as a null-labeled packet rather than an IP packet.

- **Hop 5**: The destination router PE5 successfully receives the packet, indicated by the ! symbol in the traceroute output, confirming a successful trace to the destination.

This output affirms the reliability of the TI-LFA labeled backup path from router P2 to destination PE5, ensuring loop-free path forwarding post-convergence through all intermediate routers.

With the conclusion of the single-segment FRR lab, it is time to summarize the findings in the upcoming section.

Summary

In order to explore the single-segment FRR scenario, the lab topology in *Figure 7.1* required modifications to ensure that the Zero-Segment FRR path was unavailable. Building on the previous lab setup, several strategic interfaces were shut down to achieve the desired topology suitable for single-segment FRR exploration.

The verification steps on the P2 router revealed a backup path that included an additional segment directing traffic to the P8 router before reaching the destination router PE5. This was confirmed through the label stack outputs observed in the RIB, LFIB, and FIB.

Although the lab used a small topology where visual inspection sufficed to determine the TI-LFA post-convergence path, such visual inspection is impractical in production networks with hundreds of routers and thousands of links. So, the calculation of the TI-LFA backup path was then detailed, focusing on the P-Space and extended-P-Space of P2 and the Q-Space of PE5 to identify overlapping routers. The downstream deeper node of the overlapping routers was chosen as the PQ node, or the repair node, for the backup path; in this scenario, the PQ node was P8.

While a regular traceroute identifies the best path between the source and destination through the network, the nil-fec traceroute validated that the TI-LFA backup path calculated in the control plane was also available in the data plane for use in case of failure. The nil-fec traceroute examined the backup path through each node along the path with the same label stack calculated in the control plane, and the results confirmed its availability and correctness.

This lab provided an opportunity to explore TI-LFA calculation, a foundational concept that will be beneficial in the upcoming double-segment FRR chapter.

References

The following references provide additional information and foundational specifications related to IP **Fast Reroute** (**FRR**), **Loop-Free Alternates** (**LFA**), and segment routing in MPLS networks, which are crucial for further understanding the concepts discussed in this chapter.

- *[RFC5286] Atlas, A., Ed. and A. Zinin, Ed., "Basic Specification for IP Fast Reroute: Loop-Free Alternates", RFC 5286, DOI 10.17487/RFC5286, September 2008, <https://www.rfc-editor.org/info/rfc5286>.*

- [RFC7490] Bryant, S., Filsfils, C., Previdi, S., Shand, M., and N. So, "Remote Loop-Free Alternate (LFA) Fast Reroute (FRR)", RFC 7490, DOI 10.17487/RFC7490, April 2015, <https://www.rfc-editor.org/info/rfc7490>.

- [RFC7916] Litkowski, S., Ed., Decraene, B., Filsfils, C., Raza, K., Horneffer, M., and P. Sarkar, "Operational Management of Loop-Free Alternates", RFC 7916, DOI 10.17487/RFC7916, July 2016, <https://www.rfc-editor.org/info/rfc7916>.

- [RFC8029] Kompella, K., Swallow, G., Pignataro, C., Ed., Kumar, N., Aldrin, S., and M. Chen, "Detecting Multiprotocol Label Switched (MPLS) Data-Plane Failures", RFC 8029, DOI 10.17487/RFC8029, March 2017, <https://www.rfc-editor.org/info/rfc8029>.

- [I-D.ietf-rtgwg-segment-routing-ti-lfa] Bashandy, A., Litkowski, S., Filsfils, C., Francois, P., Decraene, B., and D. Voyer, "Topology Independent Fast Reroute using Segment Routing", Work in Progress, Internet-Draft, draft-ietf-rtgwg-segment-routing-ti-lfa-13, 16 January 2024, <https://datatracker.ietf.org/doc/html/draft-ietf-rtgwg-segment-routing-ti-lfa-13>.

- [RFC5715] Shand, M. and S. Bryant, "A Framework for Loop-Free Convergence", RFC 5715, DOI 10.17487/RFC5715, January 2010, <https://www.rfc-editor.org/info/rfc5715>.

- [RFC6571] Filsfils, C., Ed., Francois, P., Ed., Shand, M., Decraene, B., Uttaro, J., Leymann, N., and M. Horneffer, "Loop-Free Alternate (LFA) Applicability in Service Provider (SP) Networks", RFC 6571, DOI 10.17487/RFC6571, June 2012, <https://www.rfc-editor.org/info/rfc6571>.

8

Lab 7 – Double-Segment FRR

As networks grow, the topologies become increasingly complex, making it challenging to provide a TI-LFA backup path using the methods explored so far in the book. This chapter covers a scenario where the backup path requires two additional segments to be encoded as labels in the backup path label stack, offering a loop-free alternate path. This increases the size of the label stack on the backup path to three, which is also the maximum number of labels supported by the Cisco IOS XR platform.

Objective

The objective of this lab is to investigate a scenario where TI-LFA provides a backup path through two additional segments, encoded as labels in the backup path label stack.

This is accomplished by the following tasks:

- Increasing the metric on the P8–P4 link from 10 to 100, which breaks the overlap between the extended P-Space of the P2 router and the Q-Space of the PE5 router.

- Verifying whether a double-segment TI-LFA backup path is provided and installed in the RIB, FIB, and LFIB.

Figure 8.1 shows the state of the topology to explore whether TI-LFA can offer a backup path or not.

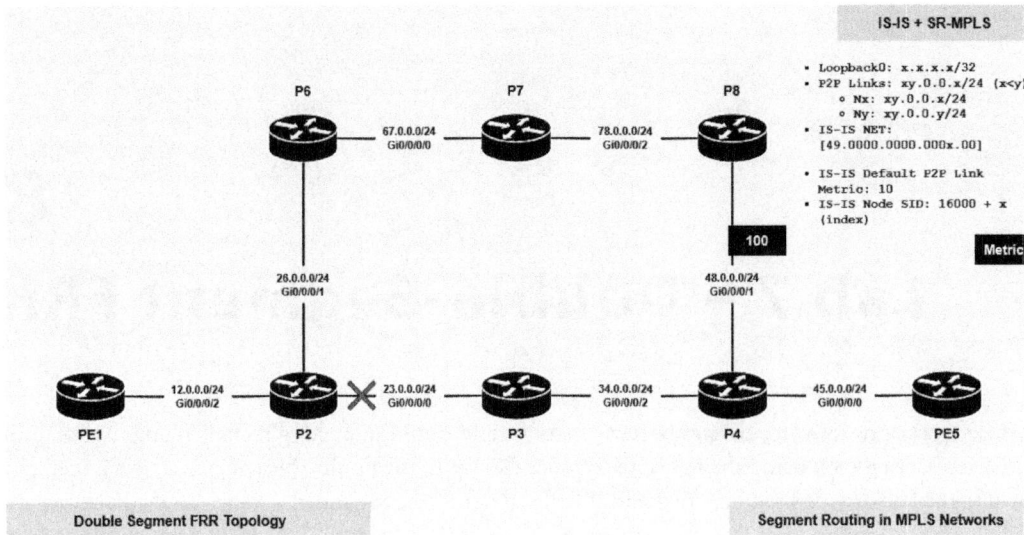

Figure 8.1 – Double-segment FRR

To achieve this state, the configuration from the next section needs to be applied accordingly.

Configuration

To establish the desired state of the topology in *Figure 8.1*, the metric on the link between the P8 and P4 routers should be increased, as illustrated here:

- Increase the metric on the P8 router toward the P4 router from 10 to 100:

P8

```
router isis IGP
interface GigabitEthernet0/0/0/1
  address-family ipv4 unicast
   metric 100
```

- Increase the metric on the P4 router toward the P8 router from 10 to 100.

P4

```
router isis IGP
interface GigabitEthernet0/0/0/1
  address-family ipv4 unicast
   metric 100
```

At first glance, it may appear that establishing a TI-LFA backup path for the primary path from P2 to PE5 is not feasible if there is a failure on the link between P2 and P3, or a failure of the P3 node.

However, this scenario and the verification in the next section provide an opportunity to understand why TI-LFA guarantees 100% backup path coverage whenever a post-convergence path is available after a failure.

Verification

In the previous chapter, the P8 router acted as the overlapping PQ node, facilitating traffic redirection from P2 to P8, which then routed it to the destination router, PE5.

However, in this case, the optimal path from the P8 router to PE5 does not pass through P4, as illustrated in the following section.

Verifying the primary path from P8 to PE5

After applying the described configuration, the primary path from P8 to PE5 explicitly passes through the P2–P3 link.

The following traceroute output illustrates that P8 utilizes the P2–P3 link to reach PE5, as the cost of passing through the P8–P4 link is higher:

```
RP/0/RP0/CPU0:P8#traceroute sr-mpls 5.5.5.5/32

Tracing MPLS Label Switched Path to 5.5.5.5/32, timeout is 2 seconds

Codes: '!' - success, 'Q' - request not sent, '.' - timeout,
   'L' - labeled output interface, 'B' - unlabeled output interface,
   'D' - DS Map mismatch, 'F' - no FEC mapping, 'f' - FEC mismatch,
   'M' - malformed request, 'm' - unsupported tlvs, 'N' - no rx label,
   'P' - no rx intf label prot, 'p' - premature termination of LSP,
   'R' - transit router, 'I' - unknown upstream index,
   'X' - unknown return code, 'x' - return code 0

Type escape sequence to abort.

  0 78.0.0.8 MRU 1500 [Labels: 16005 Exp: 0]
L 1 78.0.0.7 MRU 1500 [Labels: 16005 Exp: 0] 22 ms
L 2 67.0.0.6 MRU 1500 [Labels: 16005 Exp: 0] 18 ms
L 3 26.0.0.2 MRU 1500 [Labels: 16005 Exp: 0] 16 ms
L 4 23.0.0.3 MRU 1500 [Labels: 16005 Exp: 0] 30 ms
L 5 34.0.0.4 MRU 1500 [Labels: implicit-null Exp: 0] 25 ms
! 6 45.0.0.5 18 ms
RP/0/RP0/CPU0:P8#
```

As observed in the single-segment FRR lab, if the repair path tunnel is established to P8, then in this scenario, P8 would loop back to the P2–P3 link to reach PE5. The single-segment FRR does not offer a loop-free alternate.

The next section elaborates on how TI-LFA determines a backup path in such a scenario and the role that an additional segment plays.

Verifying double-segment backup path

The following RIB output shows that the backup path uses TI-LFA (link), indicating the default link protection method. The repair path tunnel now includes a P node, P8, and a Q node, P4. Backup Path in RIB. Interestingly, in this scenario, where the P node and the Q node are disjointed, TI-LFA has directed the backup path onto the P8 to the P4 link by encoding the additional segment as a label in the backup path, as shown here:

```
RP/0/RP0/CPU0:P2#show isis fast-reroute detail 5.5.5.5/32


L2 5.5.5.5/32 [30/115] Label: 16005, medium priority
    Installed Jan 01 07:19:18.780 for 00:00:35
      via 23.0.0.3, GigabitEthernet0/0/0/0, Label: 16005, P3, SRGB Base: 16000,
Weight: 0
        Backup path: TI-LFA (link), via 26.0.0.6, GigabitEthernet0/0/0/1 P6,
SRGB Base: 16000, Weight: 0, Metric: 140
          P node: P8.00 [8.8.8.8], Label: 16008
          Q node: P4.00 [4.4.4.4], Label: 24019
          Prefix label: 16005
          Backup-src: PE5.00
        P: No, TM: 140, LC: No, NP: No, D: No, SRLG: Yes
      src PE5.00-00, 5.5.5.5, prefix-SID index 5, R:0 N:1 P:0 E:0 V:0 L:0,
Alg:0
RP/0/RP0/CPU0:P2#
```

The P node label represents the Node-SID of P8, while the Q node label corresponds to the Adjacency-SID of the P8 router for the link toward P4. This arrangement ensures that the traffic is directed to the P8–P4 link, offering a loop-free alternate.

The routing table output now indicates two repair nodes on the backup path, 8.8.8.8 and 4.4.4.4, corresponding to the P8 and P4 nodes, respectively:

```
RP/0/RP0/CPU0:P2#show route 5.5.5.5/32


Routing entry for 5.5.5.5/32
  Known via "isis IGP", distance 115, metric 30, labeled SR, type level-2
  Installed Jan  1 07:19:18.780 for 00:01:00
  Routing Descriptor Blocks
    23.0.0.3, from 5.5.5.5, via GigabitEthernet0/0/0/0, Protected
      Route metric is 30
    26.0.0.6, from 5.5.5.5, via GigabitEthernet0/0/0/1, Backup (TI-LFA)
      Repair Node(s): 8.8.8.8, 4.4.4.4
```

```
         Route metric is 140
   No advertising protos.
RP/0/RP0/CPU0:P2#
```

The backup path label stack in the FIB should reflect the same information as the RIB, as described in the next section.

Backup path in the FIB

The FIB output displays the corresponding label stack on the backup path to implement TI-LFA:

```
RP/0/RP0/CPU0:P2#show cef 5.5.5.5/32 brief
5.5.5.5/32, version 666, labeled SR, internal 0x1000001 0x8310 (ptr 0xe7ac838)
[1], 0x600 (0xda5df48), 0xa28 (0x222f96d8)
Updated Jan  1 07:19:18.791
remote adjacency to GigabitEthernet0/0/0/0
Prefix Len 32, traffic index 0, precedence n/a, priority 1
   via 23.0.0.3/32, GigabitEthernet0/0/0/0, 10 dependencies, weight 0, class
0, protected [flags 0x400]
    path-idx 0 bkup-idx 1 NHID 0x0 [0xd673dc0 0x0]
    next hop 23.0.0.3/32
     local label 16005      labels imposed {16005}
   via 26.0.0.6/32, GigabitEthernet0/0/0/1, 11 dependencies, weight 0, class
0, backup (TI-LFA) [flags 0xb00]
    path-idx 1 NHID 0x0 [0x22032470 0x0]
    next hop 26.0.0.6/32, Repair Node(s): 8.8.8.8, 4.4.4.4
    remote adjacency
     local label 16005      labels imposed {16008 24019 16005}
RP/0/RP0/CPU0:P2#
```

The identical label stack is programmed in the LFIB as well.

Backup path in the LFIB

The LFIB label stack is programmed based on the RIB and currently holds the maximum allowable number of labels:

```
RP/0/RP0/CPU0:P2#show mpls forwarding prefix 5.5.5.5/32 detail
Local  Outgoing    Prefix           Outgoing     Next Hop        Bytes
Label  Label       or ID            Interface                    Switched
------ ----------- ---------------- ------------ --------------- ----------
--
16005  16005       SR Pfx (idx 5)   Gi0/0/0/0    23.0.0.3        360
    Updated: Jan  1 07:19:18.791
    Path Flags: 0x400 [ BKUP-IDX:1 (0xd673dc0) ]
    Version: 666, Priority: 1
    Label Stack (Top -> Bottom): { 16005 }
```

```
      NHID: 0x0, Encap-ID: N/A, Path idx: 0, Backup path idx: 1, Weight: 0
      MAC/Encaps: 4/8, MTU: 1500
      Outgoing Interface: GigabitEthernet0/0/0/0 (ifhandle 0x01000020)
      Packets Switched: 3

        16008        SR Pfx (idx
   5)   Gi0/0/0/1    26.0.0.6          0                    (!)
      Updated: Jan  1 07:19:18.791
      Path Flags: 0xb00 [  IDX:1 BKUP, NoFwd ]
      Version: 666, Priority: 1
      Label Stack (Top -> Bottom): { 16008 24019 16005 }
      NHID: 0x0, Encap-ID: N/A, Path idx: 1, Backup path idx: 0, Weight: 0
      MAC/Encaps: 4/16, MTU: 1500
      Outgoing Interface: GigabitEthernet0/0/0/1 (ifhandle 0x01000058)
      Packets Switched: 0
      (!): FRR pure backup

   Traffic-Matrix Packets/Bytes Switched: 0/0
RP/0/RP0/CPU0:P2#
```

Interestingly, the backup path entry in the LFIB for the destination, PE5, uses the Node-SID of P8, which is the downstream deeper node in the extended P-Space of P2. However, the label stack shows that the Adjacency-SID for the P8–P4 link is the second label. Therefore, once the packet arrives at P8, it consults its local LFIB to direct it toward the P4 router, despite the IGP best-cost path suggesting otherwise.

This also explains why only the Node-SIDs are globally unique throughout the network, with all routers installing those segments as labels in their FIBs and LFIBs. In contrast, the Adjacency-SID is only locally unique to the router, and only the local router installs them in its FIB and LFIB.

The backup path calculation is discussed in the next section.

Understanding backup path calculation

The TI-LFA backup path calculation method remains unchanged, but in this scenario, there is no overlapping PQ node between the extended P-Space of P2 and the Q-Space of PE5. The following explains how TI-LFA uses a combination of a Node-SID and Adjacency-SID to route traffic over a link that is not the primary path on its own.

The following topology diagram in *Figure 8.2* is relevant to the following TI-LFA calculation section:

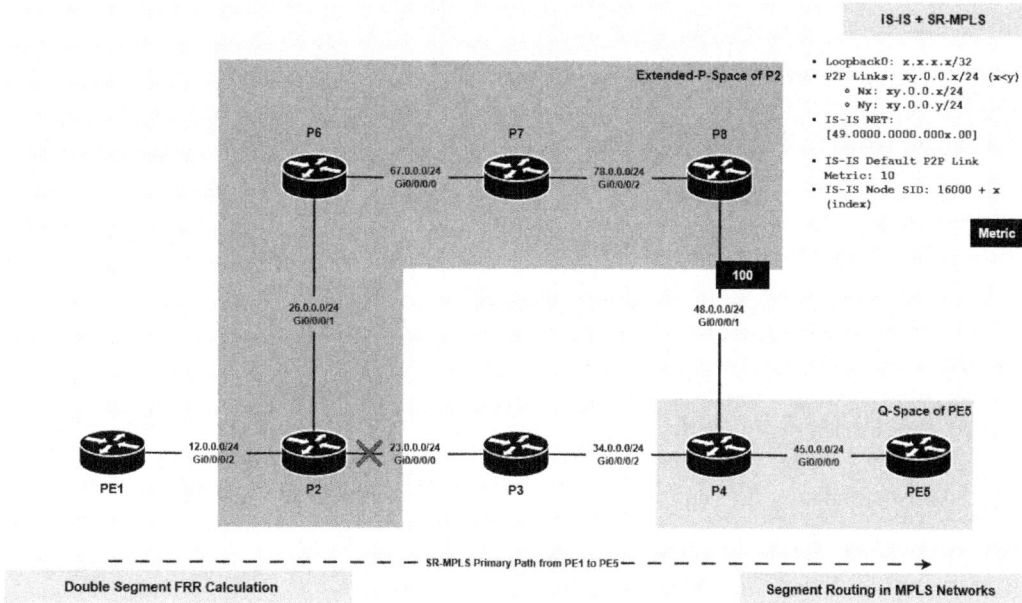

Figure 8.2 – Double-segment FRR path calculation

The following explanation details the computation and selection of the primary path and backup path.

Primary path

The primary path from P2 to PE5 goes via P3, with the end-to-end path, cost of 30 (P2 - P3 - P4 - PE5), making it the shortest path.

Backup path

The backup path is classified as TI-LFA (link) because the classic LFA for link protection inequality condition is not met here.

Classic LFA

As per RFC 5286, Inequality 1: Loop-Free Criterion is defined as, "*A neighbor N can provide a loop-free alternate (LFA) if and only if* Distance_opt(N, D) < Distance_opt(N, S) + Distance_opt(S, D)".

Here, S is used to indicate the calculating router. N_i is a neighbor of S; N is used as an abbreviation when only one neighbor is being discussed. D is the destination under consideration.

In this case, $S = P2$, $D = PE5$, and $N = P6$:

Inequality condition for P6:

```
S = P2, D = PE5 and N = P6
Distance_opt(P6, PE5) < Distance_opt(P6, P2) + Distance_opt(P2, PE5)
40 (P2 - P3 - P4 - PE5) < 10 + 30 (P3 - P4 - PE5)
```

The condition is NOT satisfied

Therefore, the Classic-LFA backup path is unavailable.

TI-LFA

TI-LFA establishes a backup path by calculating a post-convergence route and identifying P and Q nodes along it. The repair path is then guided toward the PQ or P node and then the Q node, ensuring a path to the destination. This approach guarantees 100% coverage in all situations.

In this scenario, the post-convergence path would be $P2 - P6 - P7 - P8 - P4 - PE5$ with a cost of 140.

To calculate the TI-LFA backup path for the link $P2-P3$ on the router P2, the following information will be collected:

- **P-Space of P2 along the post-convergence path**: The group of routers on the post-convergence path, reachable from $P2$ through the shortest path tree, excluding the traversal of $P2-P3$, forms the P-Space of P2. This is determined by calculating the shortest path tree, with $P2$ as the root, and removing the subtree accessed via the $P2-P3$ link (including the ECMP nodes).

  In *Figure 8.2*, these routers are P6, P7, and P8.

- **Extended P-Space of P2 along the post-convergence path**: The extended P-Space of P2 is the combination of P2's P-Space and the P-Space of P2's neighbors along the post-convergence path. This can be determined by calculating the shortest path tree, rooted at each of P2's neighbors, and removing the subtree accessed via the $P2-P3$ link (including the ECMP nodes).

  The P-Space of P6 comprises P7 and P8.

  Therefore, the extended P-Space of P2 along the post-convergence path includes P6, P7, and P8.

- **Q-Space of PE5 along the post-convergence path**: The group of routers along the post-convergence path, through which the PE5 router can be reached via the shortest path tree without traversing the $P2-P3$ link, forms the Q-Space of PE5. This is achieved by computing a reverse shortest path tree, rooted at PE5, and removing the subtree reached via the $P2-P3$ links (including the ECMP nodes).

  The Q-Space of PE5 comprises P4 only (as P3 is not along the post-convergence path).

The TI-LFA repair path is a sequence of segments (i.e., a stack of labels) along the post-convergence path on the outgoing interface. This path is determined by the intersection of the extended P-Space of P2 and the Q-Space of PE5.

There is no overlap between the extended P-Space and Q-Space in this scenario. The deeper downstream P node is P8, and the Q node adjacent to it is P4.

The endpoint for the repair tunnel must be P4 along the post-convergence path. To achieve this, P2 encodes additional information in the label stack that instructs P8 to use the adjacency toward P4, providing additional instructions to ensure a loop-free alternate.

This is evident in the output where the outgoing label on the backup path from the **Point of Local Repair (PLR)**, P2, to the destination, PE5, is the Node-SID of P8 and the Adjacency-SID on P8 to P4.

If there is a P2–P3 link failure, the P2 router redirects the traffic to the backup path, leading it through P8 and then P4. The path from P2 to P8 does not involve the P2–P3 link, and similarly, the path from P4 to PE5 avoids the P2–P3 link. Upon receiving the traffic, P8 observes an additional label, the Adjacency-SID of P4, and refers to its LFIB to forward it. The subsequent path from P4 to PE5 also does not use the P2–P3 link. Thus, the TI-LFA backup path ensures loop-free forwarding.

Using the Adjacency-SID labels, TI-LFA directs the route between non-overlapping P- and Q-Space routers, ensuring that a post-convergence path also serves as the backup path for fast rerouting of traffic if there is a failure.

Verifying the backup path traceroute

The essential components necessary for the nil-FEC traceroute, as derived from the RIB, FIB, or LFIB, are as follows:

- **Label stack (top to bottom)**: 16008 24019 16005, directing traffic to P8, then P8–P4, and finally, PE5.

- **Outgoing interface**: GigabitEthernet0/0/0/1, toward P6.

- **Next-hop IP address of the outgoing interface**: 26.0.0.6, to P6.

> **Note**
>
> The Adjacency-SID label is allocated from the dynamic label space, so each iteration of this lab might display a different label for everyone.

The nil-fec traceroute for the backup path from P2 to PE5, as per the preceding information, is displayed as follows:

```
RP/0/RP0/CPU0:P2#traceroute sr-mpls nil-fec labels 16008,24019,16005 output
interface GigabitEthernet0/0/0/1 nexthop 26.0.0.6

Tracing MPLS Label Switched Path with Nil FEC with labels [16008,24019,16005],
timeout is 2 seconds
```

```
Codes: '!' - success, 'Q' - request not sent, '.' - timeout,
  'L' - labeled output interface, 'B' - unlabeled output interface,
  'D' - DS Map mismatch, 'F' - no FEC mapping, 'f' - FEC mismatch,
  'M' - malformed request, 'm' - unsupported tlvs, 'N' - no rx label,
  'P' - no rx intf label prot, 'p' - premature termination of LSP,
  'R' - transit router, 'I' - unknown upstream index,
  'X' - unknown return code, 'x' - return code 0

Type escape sequence to abort.

  0 26.0.0.2 MRU 1500 [Labels: 16008/24019/16005/explicit-null Exp: 0/0/0/0]
L 1 26.0.0.6 MRU 1500 [Labels: 16008/24019/16005/explicit-null Exp: 0/0/0/0]
14 ms
L 2 67.0.0.7 MRU 1500 [Labels: implicit-null/24019/16005/explicit-null Exp:
0/0/0/0] 19 ms
L 3 78.0.0.8 MRU 1500 [Labels: implicit-null/16005/explicit-null Exp: 0/0/0/0]
11 ms
L 4 48.0.0.4 MRU 1500 [Labels: implicit-null/explicit-null Exp: 0/0] 19 ms
! 5 45.0.0.5 21 ms
RP/0/RP0/CPU0:P2#
```

The following topology diagram in *Figure 8.3* shows the primary and backup paths:

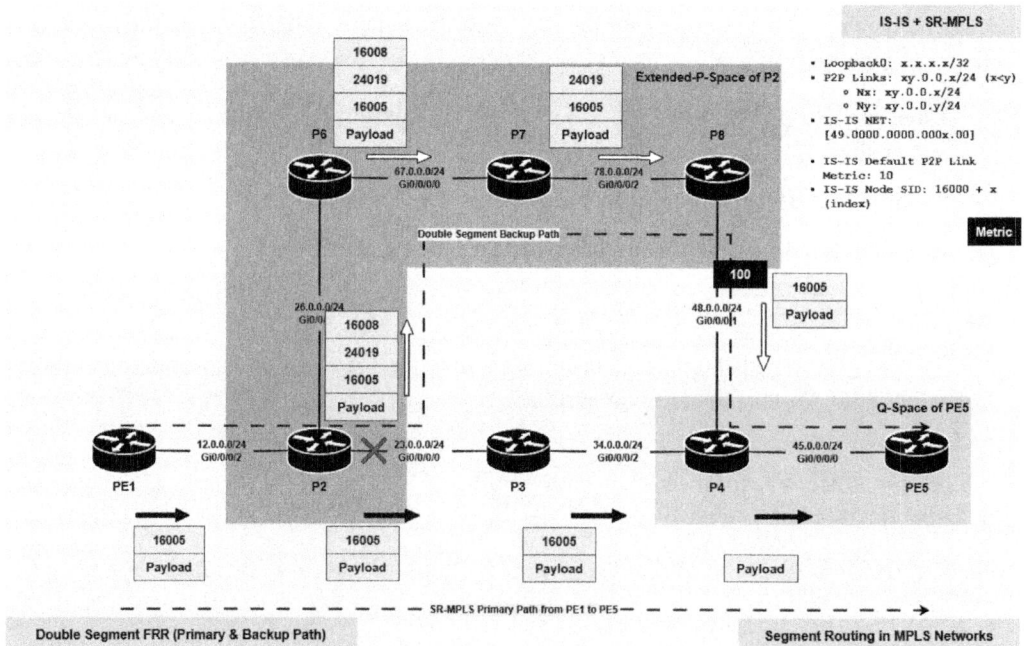

Figure 8.3 – Double-segment FRR (primary and backup paths)

Each of the hops and the corresponding label stack for the TI-LFA MPLS backup path traceroute, from the P2 router to the PE5 router, is explained as follows:

1. **Hop 0**: The traceroute originates from the P2 router, with an MPLS label stack of 16008/24019/16005/explicit-null. The inclusion of "explicit-null" ensures that the penultimate router does not remove the MPLS labels prematurely.

2. **Hop 1**: The trace exits router P6, which inspects the top MPLS label, consults its **LFIB** (**Label Forwarding Information Base**), swaps the top label with the outgoing label (16008), and forwards the packet based on the LFIB's next-hop and outgoing interface information.

3. **Hop 2**: The trace proceeds through the P7 router, which examines the top MPLS label, checks its LFIB, and determines whether it is the penultimate router for P8. It pops the top label (hence, implicit-null) and forwards the packet along the MPLS label-switched path.

4. **Hop 3**: The trace moves through the P8 router, receiving the top MPLS label as 24019. P8 consults the LFIB to find out that it is the adjacency label for the link toward P4. Since it is the penultimate hop to P4, it pops the Adjacency-SID label (hence, implicit-null) and forwards the packet along the MPLS label-switched path.

5. **Hop 4**: The trace passes through router P4, which serves as the penultimate router for PE5. The MPLS label stack includes implicit-null and explicit-null. This configuration ensures that P4 forwards the packet as a null-labeled packet rather than an IP packet.

6. **Hop 5**: The destination router PE5 successfully receives the packet, indicated by the ! symbol in the traceroute output, confirming a successful trace to the destination.

7. In SR-MPLS networks, routers install only their own Adjacency-SID labels in their data planes (the FIB and LFIB). Thus, router P2 installs its dynamically allocated Adjacency-SID labels for its IS-IS neighbors. However, routers also propagate this information via link state packet updates across a network.

Router P2 acquired the Adjacency-SID of router P8 via IS-IS and utilized it to program the label stack for the backup path.

Next, the Adjacency-SID of P8 will be verified.

The PLR router, P2, uses the non-FRR Adjacency-SID from P8 to P4, which it learned from the IS-IS LSPDU advertised by P8, as shown here:

```
RP/0/RP0/CPU0:P8#show isis adjacency detail systemid P4

IS-IS IGP Level-2 adjacencies:
System Id       Interface             SNPA         State Hold Changed   NSF
IPv4 IPv6
                                                         BFD   BFD
P4              Gi0/0/0/1             *PtoP*        Up    25    01:21:52 Yes
None None
   Area Address:           49
   Neighbor IPv4 Address:  48.0.0.4*
```

```
   Adjacency SID:            24018 (protected)
    Backup label stack:      [16004]
    Backup stack size:       1
    Backup interface:        Gi0/0/0/2
    Backup nexthop:          78.0.0.7
    Backup node address:     4.4.4.4
    Non-FRR Adjacency SID:   24019
    Topology:                IPv4 Unicast
    BFD Status:              BFD Not Required, Neighbor Useable

 Total adjacency count: 1
 RP/0/RP0/CPU0:P8#
```

Over the backup path, whether during fast reroute or `nil-fec` traceroute, P8 receives the non-FRR Adjacency-SID as the outermost label. Since P4 is its adjacent node, the LFIB operation once more involves popping the label, illustrated as follows. This explains why the third hop shows that P4 only receives the Node-SID of PE5:

```
RP/0/RP0/CPU0:P8#show mpls forwarding labels 24019 detail
Local   Outgoing   Prefix              Outgoing      Next Hop         Bytes
Label   Label      or ID               Interface                      Switched
------  ---------- ------------------- ------------- ---------------- ----------
--
24019  Pop         SR Adj (idx 3)      Gi0/0/0/1     48.0.0.4         260
     Updated: Jan  1 06:54:03.953
     Version: 199, Priority: 1
     Label Stack (Top -> Bottom): { Imp-Null }
     NHID: 0x0, Encap-ID: N/A, Path idx: 0, Backup path idx: 0, Weight: 0
     MAC/Encaps: 4/4, MTU: 1500
     Outgoing Interface: GigabitEthernet0/0/0/1 (ifhandle 0x01000050)
     Packets Switched: 2

RP/0/RP0/CPU0:P8#
```

This traceroute confirms that if a post-convergence path is available for a link failure in the network, TI-LFA will provide that path as a backup. This ensures 100% backup path coverage.

Restoring the topology

Restore the topology as follows.

- Reduce the metric on the P8 router toward the P4 router from 100 to 10:

  P8

  ```
  router isis IGP
  interface GigabitEthernet0/0/0/1
  ```

```
  address-family ipv4 unicast
    metric 10
```

- Increase the metric on the P4 router toward the P8 router from 100 to 10.

P4

```
  router isis IGP
  interface GigabitEthernet0/0/0/1
    address-family ipv4 unicast
      metric 10
```

That concludes the double-segment FRR lab, and we now have the topology restored for the next chapter.

Summary

In this chapter, we modified the network topology to render single-segment FRR inapplicable. The adjustment necessitated that TI-LFA accommodates additional information and calculations, ensuring that the post-convergence path also serves as a backup path if there is a failure.

Following an increase in metric between the P8 and P4 routers from 10 to 100, the primary path from P8 to PE5 shifted to utilize the P2–P3 link, as observed in the traceroute of the primary path. Depending on whether the backup path carried the Node-SID of P8 or P4, distinct routing behaviors were noted – if using P8's Node-SID, traffic would route back via P2, whereas with P4's Node-SID, the route would proceed directly through P2–P3 due to the higher cost via P8–P4.

Visually, it became apparent that if there was a P2–P3 link failure, the post-convergence path would necessarily traverse P8 and P4 to reach PE5. Analysis of the backup path output across RIB, FIB, and LFIB revealed an intriguing detail – it utilized the Adjacency-SID of the P8–P4 link alongside P8's Node-SID in the label stack, instructing P8 to direct traffic to P4 over this adjacency.

Verification via the backup path traceroute confirmed functionality within MPLS networks, where **label-switched paths** (**LSPs**) rely on the LFIB for label lookup. The label stack's inclusion of P8's node-SID and P8–P4's Adjacency-SID ensured a clear path, from P4 onward to the destination router, PE5.

The chapter further delved into TI-LFA backup path calculations within the double-segment FRR scenario, emphasizing the role of Adjacency-SIDs in SR-MPLS networks. We clarified that while IS-IS advertises both SRGB alongside Node-SID indexes and Adjacency-SIDs as absolute values, Adjacency-SIDs are locally significant and only installed in originating router's LFIBs, unlike globally significant node-SIDs, which are installed across all routers in the SR-MPLS network. Remote routes use the Node-SID of the Adjacency-SID originator and the Adjacency-SID itself as the next label in the stack.

Ultimately, the topology was restored to align with the objectives set for the next chapter.

Before continuing with additional TI-LFA scenarios to enhance our understanding of segment routing, we will introduce Microloop avoidance in SR-MPLS with TI-LFA networks in the next chapter.

References

The following references provide additional information and foundational specifications related to IP **Fast Reroute** (**FRR**), **Loop-Free Alternates** (**LFA**), and segment routing in MPLS networks, which are crucial for further understanding the concepts discussed in this chapter.

- *[RFC5286] Atlas, A., Ed. and A. Zinin, Ed., "Basic Specification for IP Fast Reroute: Loop-Free Alternates", RFC 5286, DOI 10.17487/RFC5286, September 2008, <https://www.rfc-editor.org/info/rfc5286>.*

- *[RFC7490] Bryant, S., Filsfils, C., Previdi, S., Shand, M., and N. So, "Remote Loop-Free Alternate (LFA) Fast Reroute (FRR)", RFC 7490, DOI 10.17487/RFC7490, April 2015, <https://www.rfc-editor.org/info/rfc7490>.*

- *[RFC7916] Litkowski, S., Ed., Decraene, B., Filsfils, C., Raza, K., Horneffer, M., and P. Sarkar, "Operational Management of Loop-Free Alternates", RFC 7916, DOI 10.17487/RFC7916, July 2016, <https://www.rfc-editor.org/info/rfc7916>.*

- *[RFC8029] Kompella, K., Swallow, G., Pignataro, C., Ed., Kumar, N., Aldrin, S., and M. Chen, "Detecting Multiprotocol Label Switched (MPLS) Data-Plane Failures", RFC 8029, DOI 10.17487/RFC8029, March 2017, <https://www.rfc-editor.org/info/rfc8029>.*

- *[I-D.ietf-rtgwg-segment-routing-ti-lfa] Bashandy, A., Litkowski, S., Filsfils, C., Francois, P., Decraene, B., and D. Voyer, "Topology Independent Fast Reroute using Segment Routing", Work in Progress, Internet-Draft, draft-ietf-rtgwg-segment-routing-ti-lfa-13, 16 January 2024, <https://datatracker.ietf.org/doc/html/draft-ietf-rtgwg-segment-routing-ti-lfa-13>.*

- *[RFC5715] Shand, M. and S. Bryant, "A Framework for Loop-Free Convergence", RFC 5715, DOI 10.17487/RFC5715, January 2010, <https://www.rfc-editor.org/info/rfc5715>.*

- *[RFC6571] Filsfils, C., Ed., Francois, P., Ed., Shand, M., Decraene, B., Uttaro, J., Leymann, N., and M. Horneffer, "Loop-Free Alternate (LFA) Applicability in Service Provider (SP) Networks", RFC 6571, DOI 10.17487/RFC6571, June 2012, <https://www.rfc-editor.org/info/rfc6571>.*

9

Lab 8 – Microloop Avoidance

This chapter provides a refreshing break from the **Topology-Independent Loop-Free Alternate (TI-LFA)** scenarios, which will resume in the upcoming chapters. It is well-suited to the current topology, requiring minimal changes from the previous lab and seamlessly transitioning to the next chapter. The TI-LFA labs have thoroughly covered the calculation of backup paths and the protection of traffic during failures by rerouting it to the post-convergence backup path.

However, what happens if a network topology change causes the primary path to potentially loop?

All routers running a link-state routing protocol, such as IS-IS, populate the **Link State Database (LSDB)** and independently run the **Shortest Path First (SPF)** algorithm to create a uniform topology map. This process results in routers converging at different times. During a topology change, some routers may have an updated map while others are still processing the SPF algorithm to arrive at the same new map. This delay in protocol convergence leads to microloops in the network, which can drop traffic on the primary path.

Fortunately, there is a solution. By applying the same calculation methods used in the TI-LFA labs, it is possible to protect the primary path itself by tunneling it to the repair node in such scenarios.

Objective

The objective of this lab is to explore a scenario where routers anticipate a microloop in the network due to a change in topology and employ necessary microloop avoidance mechanisms to maintain a continuous flow of traffic, which would otherwise be impacted by the loops. This is accomplished through the following tasks:

1. Adjust the configuration to align with the depicted topology in the diagram.

2. Apply the configuration for microloop avoidance on P2.

3. Initiate a change in the network topology to trigger microloop avoidance.

4. Verify that a loop-free explicit SR-MPLS primary path from P2 to PE5 is computed and added into the **Routing Information Base (RIB)**, **Forwarding Information Base (FIB)**, and **Label Forwarding Information Base (LFIB)**.

The topology shown in *Figure 9.1* is used for the purpose of this lab.

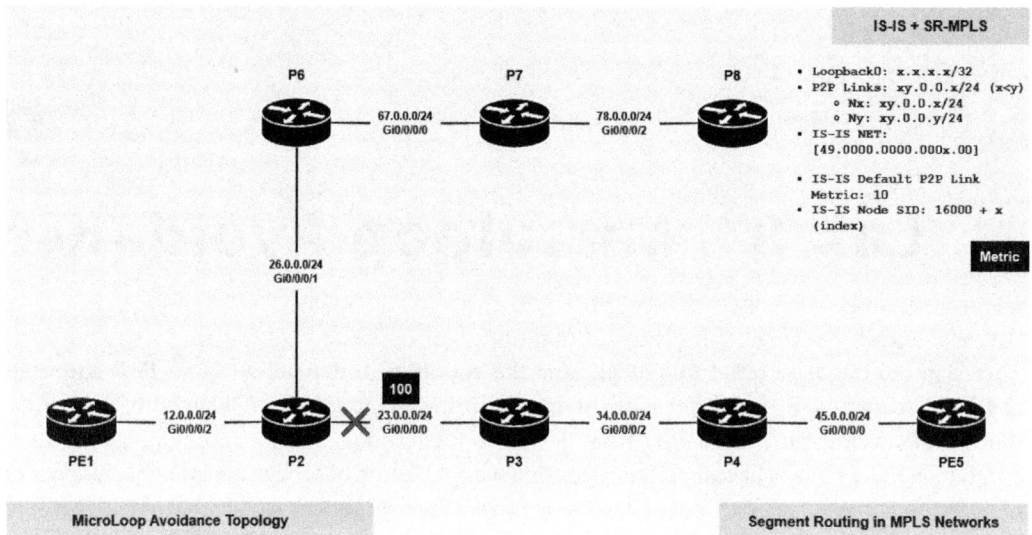

Figure 9.1 – Microloop avoidance

The next section covers the configuration required to achieve this state as shown in *Figure 9.1*.

Preparation

Increase the metric on the P2 router toward P3 on the P2-P3 link and shut down the P8-P4 interface on the P8 router. This adjustment causes the P8 router to use the P2-P3 link to reach the PE5 router.

On router P2, increase the metric toward router P3 from 10 to 100:

P2

```
router isis IGP
interface GigabitEthernet0/0/0/0
  address-family ipv4 unicast
  metric 100
```

Additionally, shut down the interface on P8 toward P4:

P8

```
router isis IGP
interface GigabitEthernet0/0/0/1
  shutdown
```

This configuration should result in no backup path from P2 to PE5, as shown below.

Verification

The next section verifies that the desired state of the topology has been achieved.

Verify the primary path from P2 to PE5

The following output verifies that the path from P2 to PE5 is via P3.

Primary path in the RIB

Even though the P2-P3 link has a cost of 100, since there is no other route to the destination PE5, there is no backup path in the RIB, FIB, or LFIB.

Notice No FRR backup in the following output, emphasizing that there is no backup route in case the primary path fails:

```
RP/0/RP0/CPU0:P2#show isis fast-reroute detail 5.5.5.5/32


L2 5.5.5.5/32 [120/115] Label: 16005, medium priority
    Installed Jan 01 07:26:46.497 for 00:00:16
       via 23.0.0.3, GigabitEthernet0/0/0/0, Label: 16005, P3, SRGB Base: 16000,
Weight: 0
       No FRR backup
       src PE5.00-00, 5.5.5.5, prefix-SID index 5, R:0 N:1 P:0 E:0 V:0 L:0,
Alg:0
RP/0/RP0/CPU0:P2#
```

Notice that the metric to reach PE5 from P2 is 120; this will be important to compare when micro-loop avoidance is in progress later in the chapter.

Hence, the routing table output shows no backup path. There is only a primary path to PE5 (16005) via the P2-P3 link (23.0.0.3):

```
RP/0/RP0/CPU0:P2#show route 5.5.5.5/32 detail


Routing entry for 5.5.5.5/32
  Known via "isis IGP", distance 115, metric 120, labeled SR, type level-2
  Installed Jan  1 07:26:46.497 for 00:00:35
  Routing Descriptor Blocks
    23.0.0.3, from 5.5.5.5, via GigabitEthernet0/0/0/0
      Route metric is 120
      Label: 0x3e85 (16005)
      Tunnel ID: None
      Binding Label: None
      Extended communities count: 0
      Path id:1        Path ref count:0
      NHID:0x7(Ref:10)
```

```
  Route version is 0x30 (48)
  Local Label: 0x3e85 (16005)
  IP Precedence: Not Set
  QoS Group ID: Not Set
  Flow-tag: Not Set
  Fwd-class: Not Set
  Route Priority: RIB_PRIORITY_NON_RECURSIVE_MEDIUM (7) SVD Type RIB_SVD_TYPE_
LOCAL
  Download Priority 1, Download Version 726
  No advertising protos.
RP/0/RP0/CPU0:P2#
```

The preceding detailed output shows no repair node, no backup path, and so on.

Primary path in the FIB

The same information is downloaded into the FIB:

```
RP/0/RP0/CPU0:P2#show cef 5.5.5.5/32 brief
5.5.5.5/32, version 726, labeled SR, internal 0x1000001 0x8310 (ptr 0xe7ac838)
[1], 0x600 (0xda5df48), 0xa28 (0xebfe3c8)
Updated Jan  1 07:26:46.501
remote adjacency to GigabitEthernet0/0/0/0
Prefix Len 32, traffic index 0, precedence n/a, priority 1
   via 23.0.0.3/32, GigabitEthernet0/0/0/0, 8 dependencies, weight 0, class 0
[flags 0x0]
    path-idx 0 NHID 0x0 [0x22032330 0x0]
    next hop 23.0.0.3/32
    remote adjacency
     local label 16005     labels imposed {16005}
RP/0/RP0/CPU0:P2#
```

Primary path in the LFIB

The LFIB also downloads the same information from the RIB:

```
RP/0/RP0/CPU0:P2#show mpls forwarding prefix 5.5.5.5/32
Local   Outgoing    Prefix              Outgoing    Next Hop        Bytes
Label   Label       or ID               Interface                   Switched
------  ----------- ------------------- ----------- --------------- ----------
--
16005   16005       SR Pfx (idx 5)      Gi0/0/0/0   23.0.0.3        0
RP/0/RP0/CPU0:P2#
```

If the P2-P3 link goes down, there is no other path for P2 to reach PE5. Therefore, there wouldn't be a post-convergence path, and hence, no TI-LFA backup path.

When the P8-P4 link comes up, the path from P2 to PE5 via the P8-P4 link will have a better end-to-end cost compared to the path via the P2-P3 link, due to the higher cost of the latter. The IS-IS routing protocol will flood the link-state information of the P8-P4 link throughout the network. Once the routers have updated their LSDB, they will run the SPF algorithm. Routers P2, P6, P7, and P8 will then prefer the P8-P4 link to reach the PE5 router due to its lower cost compared to the higher cost of the path via the P2-P3 link.

However, routers P2, P6, P7, and P8 could reach that conclusion at different times. It is possible that while P2 finishes running the SPF algorithm and switches the primary path next hop from P3 to P6, P6 or any other router along the path might still be running the SPF and routing traffic based on older information, directing it back toward P2. This causes a loop. The issue eventually resolves itself when convergence is completed, but during that brief period, there would be traffic loss in the network due to the microloops.

Now, before we move on to configuration, the next section offers a brief introduction to microloop avoidance in a nutshell.

Microloop avoidance in SR-MPLS networks

During topology changes, the network undergoes a process of flooding link state information across its nodes. Each router independently employs the shortest path algorithm to construct its own representation of the topology. However, this decentralized approach may introduce microloops in the network as the convergence of internal gateway protocols might not be synchronized among all routers.

To address this issue, SR-MPLS networks implement microloop avoidance configurations, as outlined in the IETF proposal, **draft-bashandy-rtgwg-segment-routing-uloop**. In cases where a source router lacks confidence in the loop-free nature of the post-convergence path to a destination, it can establish an explicit path without generating any state along that route. This explicit path is maintained briefly, providing sufficient time for the network to converge. Subsequently, the post-convergence path is installed in the RIB, FIB, and LFIB.

The stage is set to introduce additional configurations for implementing microloop avoidance in the network in the next section.

Configuration

Once a topology without a backup path from P2 to PE5 is achieved, apply the commands shown on router P2 to enable microloop avoidance.

The following commands enable microloop avoidance for segment routing in the IS-IS protocol and configure an RIB update delay of 60,000 milliseconds or 60 seconds:

P2

```
router isis IGP
address-family ipv4 unicast
```

```
microloop avoidance segment-routing
microloop avoidance rib-update-delay 60000
```

This represents the maximum configurable time, assuming that **Interior Gateway Protocol (IGP)** convergence is expected to be completed across all routers within the update delay.

The loop-free explicit path remains active during this period, following which the post-convergence path takes over.

The following output confirms that microloop avoidance is enabled on router P2:

```
RP/0/RP0/CPU0:P2#show isis protocol | utility egrep -A2 Micro
       Microloop avoidance: Enabled
          Configuration: Type: Segment routing, RIB update delay: 60000 msec
      No protocols redistributed
RP/0/RP0/CPU0:P2#
```

Once triggered, it will remain in effect for 60 seconds before reverting to the post-convergence path.

> **Note**
>
> For the purpose of this lab scenario, the microloop avoidance is configured only on the P2 router. In production networks, it should be enabled on all the SR-MPLS routers and `rib-update-delay` should be optimized according to the size of the production network.

The next step is to trigger a change in the topology, as depicted in the following section.

Trigger microloop avoidance

Bring up the link between P8 and P4, triggering a change in the topology and creating a potential loop scenario.

Activate the interface toward P4 in the IS-IS context:

P8

```
router isis IGP
interface GigabitEthernet0/0/0/1
  no shutdown
```

Before this, the P8 to PE5 path utilized the P2-P3 link. After bringing up the P8-P4 link, P2 might start forwarding traffic to P8. However, if P8 or any other router along the path, such as P6 or P7, has not completed the convergence process, there is a possibility of routing the traffic back toward P2, resulting in undesired microloops.

This is confirmed in the next section.

Verification

There are only 60,000 milliseconds or 60 seconds during which the microloop avoidance mechanism will be in effect. After that, the post-convergence path will take over. Therefore, the following verification section needs to be completed quickly.

Verifying the microloop avoidance primary path from P2 to PE5

As expected, the change in topology has initiated microloop avoidance on router P2. The following output confirms a shift to the active state, detecting a link-up event on the P8-P4 link. The duration specified represents the time during which microloop avoidance ensures loop-free SR-MPLS forwarding in the RIB, FIB, and LFIB:

```
RP/0/RP0/CPU0:P2#show isis protocol | utility egrep -A2 Micro
        Microloop avoidance: Enabled
          Configuration: Type: Segment routing, RIB update delay: 60000 msec
          State: Active, Duration: 7822 ms, Event Link up, Near: P8.00 Far:
P4.00
RP/0/RP0/CPU0:P2#
```

The selection of a 60-second RIB update delay in this lab serves the purpose of allowing ample time to capture outputs that depict the explicit loop-free SR-MPLS path on the P2 router.

To prevent potential problems with microloops along the post-convergence path, it's important to tunnel the loop-free path explicitly through the P8 router. Additional instructions in the form label stack are provided to ensure it uses the connection to P4, enabling continuous forwarding until the network's internal routing processes are completed.

> **Note**
> Unlike TI-LFA, this is not a backup path but rather a temporary primary path.

The following RIB output shows that P2 has an explicit route to PE5 via P8 and P4, with more than one label in the label stack of the primary path. This differs from before when only the destination router Node-SID was the sole label in the label stack of the primary path.

Explicit primary path in the RIB

The following output shows the primary path along with an explicit path from P2 to PE5:

```
RP/0/RP0/CPU0:P2#show isis fast-reroute detail 5.5.5.5/32

L2 5.5.5.5/32 [50/115] Label: 16005, medium priority
   Installed Jan 01 07:31:55.368 for 00:00:28
      via 26.0.0.6, GigabitEthernet0/0/0/1, P6, SRGB Base: 16000, Weight: 0
```

```
      No FRR backup
    exp 26.0.0.6, GigabitEthernet0/0/0/1, P6, SRGB Base: 16000, Weight: 0
    via explicit path
       P node: P8.00 [8.8.8.8], Label: 16008
       Q node: P4.00 [4.4.4.4], Label: 24019
       Prefix label: 16005
    src PE5.00-00, 5.5.5.5, prefix-SID index 5, R:0 N:1 P:0 E:0 V:0 L:0,
Alg:0
RP/0/RP0/CPU0:P2#
```

Here are a few observations from the preceding output:

- The primary path, which previously used P3 as the next hop, has now switched to P6 as the next hop. The end-to-end cost was 120 earlier; now it is 50, and the lower cost is preferred.

- Due to a triggered update in the topology, microloop avoidance on router P2 has become active. Upon calculating the new primary path, due to the microloop avoidance configuration, P2 assumed that other hops along the new path might not have converged. Therefore, it created an explicit path to the destination, which includes instructions to tunnel the traffic to P8 and then direct it to router P4. This ensures that routers along the path do not loop traffic back to P2.

- There is no FRR backup, emphasizing that this is the primary path.

The following routing table output also reflects the same:

```
RP/0/RP0/CPU0:P2#show route 5.5.5.5/32 detail

Routing entry for 5.5.5.5/32
  Known via "isis IGP", distance 115, metric 50, labeled SR, type level-2
  Installed Jan  1 07:31:55.368 for 00:00:38
  Routing Descriptor Blocks
    26.0.0.6, from 5.5.5.5, via GigabitEthernet0/0/0/1
      Route metric is 50
      Labels: 0x3e88 0x5dd3 0x3e85 (16008 24019 16005)
      Tunnel ID: None
      Binding Label: None
      Extended communities count: 0
      Path id:1      Path ref count:0
      NHID:0x6(Ref:16)
  Route version is 0x32 (50)
  Local Label: 0x3e85 (16005)
  IP Precedence: Not Set
  QoS Group ID: Not Set
  Flow-tag: Not Set
  Fwd-class: Not Set
  Route Priority: RIB_PRIORITY_NON_RECURSIVE_MEDIUM (7) SVD Type RIB_SVD_TYPE_
LOCAL
```

```
  Download Priority 1, Download Version 747
  No advertising protos.
RP/0/RP0/CPU0:P2#
```

The label stack clearly shows that the path from P2 to PE5 is using the Node-SID of router P8, then the Adjacency-SID for the P8-P4 link, and finally, the Node-SID of the destination router PE5.

The explicit path is also verified in the data plane outputs of both the FIB and LFIB.

Explicit primary path in the FIB

The label stack in the FIB confirms that the explicit path is installed in the FIB:

```
RP/0/RP0/CPU0:P2#show cef 5.5.5.5/32 brief
5.5.5.5/32, version 747, labeled SR, internal 0x1000001 0x8310 (ptr 0xe7ac838)
[1], 0x600 (0xda5df48), 0xa28 (0xebfeb98)
Updated Jan  1 07:31:55.377
remote adjacency to GigabitEthernet0/0/0/1
Prefix Len 32, traffic index 0, precedence n/a, priority 1
   via 26.0.0.6/32, GigabitEthernet0/0/0/1, 10 dependencies, weight 0, class 0
[flags 0x0]
    path-idx 0 NHID 0x0 [0x22032470 0x0]
    next hop 26.0.0.6/32
    remote adjacency
    local label 16005        labels imposed {16008 24019 16005}
RP/0/RP0/CPU0:P2#
```

The primary path is tunneled to router P8 and then directed through P8-P4 toward the destination PE5.

Explicit primary path in the LFIB

The LFIB downloaded the information accordingly from the RIB to reflect the explicit path in the labeled forwarding table:

```
RP/0/RP0/CPU0:P2#show mpls forwarding prefix 5.5.5.5/32 detail
Local  Outgoing    Prefix             Outgoing     Next Hop         Bytes
Label  Label       or ID              Interface                     Switched
------ ----------- ------------------ ------------ ---------------- ----------
--
16005  16008       SR Pfx (idx 5)     Gi0/0/0/1    26.0.0.6         0
    Updated: Jan  1 07:31:55.377
    Version: 747, Priority: 1
    Label Stack (Top -> Bottom): { 16008 24019 16005 }
    NHID: 0x0, Encap-ID: N/A, Path idx: 0, Backup path idx: 0, Weight: 0
    MAC/Encaps: 4/16, MTU: 1500
    Outgoing Interface: GigabitEthernet0/0/0/1 (ifhandle 0x01000058)
    Packets Switched: 0
```

```
Traffic-Matrix Packets/Bytes Switched: 0/0
RP/0/RP0/CPU0:P2#
```

This explicit path will remain microloop in effect in both the control plane and the data plane for the duration of the RIB update delay, which is configured to be 60 seconds, as verified in the next section.

The next section provides an overview of the explicit path calculation.

Understanding the explicit primary path calculation

The following diagram depicts the explicit primary path to avoid microloops in the network.

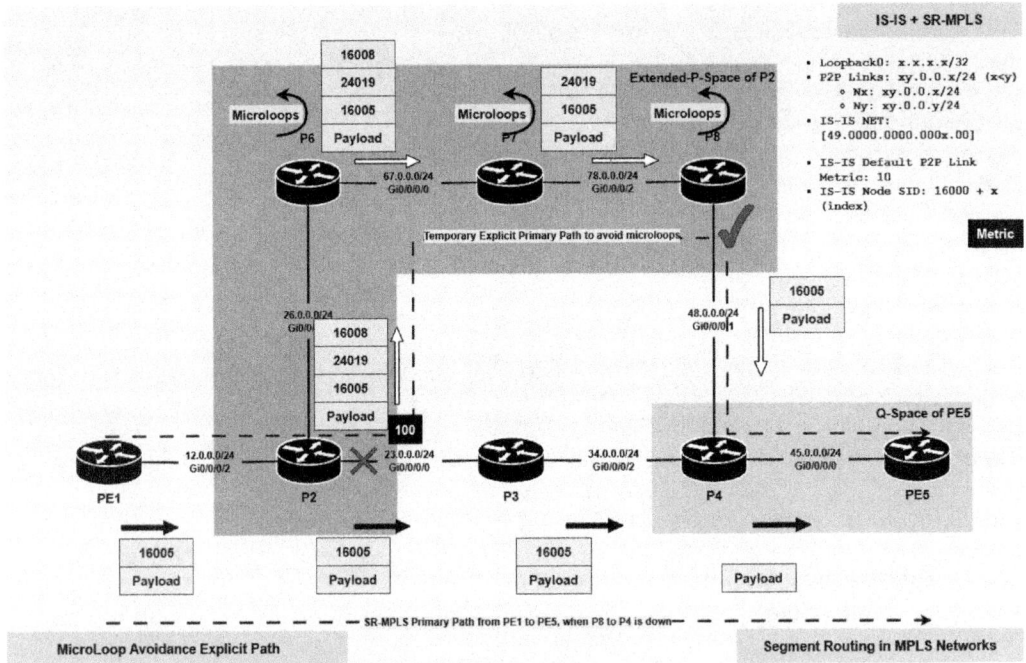

Figure 9.2 – Microloop Avoidance (explicit loop-free path)

The following explanation for the calculation of the explicit path is similar to that of the TI-LFA backup path, but this one applies to the primary path:

- **Explicit post-convergence primary path**: In this scenario, the post-convergence path from P2 to PE5 would be P2 - P6 - P7 - P8 - P4 - PE5 with a cost of 50.

- **Extended P-Space of P2**: The extended P-Space of P2 comprises P6, P7, and P8. This calculation is similar to the one seen in *Chapter 7*.

- **Q-Space of PE5**: The Q-Space of PE5 comprises P4 only. The link between P8 and P4 is in the post-convergence path.

There is no overlap between the extended P-Space and Q-Space in this scenario. The deeper downstream P node is P8, and the adjacent Q node to it is P4.

The endpoint for the repair tunnel needs to be P4 along the post-convergence path. To achieve this, router P2 includes specific labels (P8 Node-SID and P8-P4 Adjacency-SID) that instruct P8 to use the connection to P4.

Once the RIB update delay of 60,000 milliseconds (60 seconds) expires, the post-convergence path without any tunnel replaces the explicit path.

This is verified in the next section.

Verifying the post-convergence primary path from P2 to PE5

Following the 60-second RIB update delay, microloop avoidance is deactivated, and the post-convergence path is installed until another topology change occurs, potentially introducing microloops.

The state information is no longer present, as confirmed here:

```
RP/0/RP0/CPU0:P2#show isis protocol | utility egrep -A2 Micro
        Microloop avoidance: Enabled
            Configuration: Type: Segment routing, RIB update delay: 60000 msec
        No protocols redistributed
RP/0/RP0/CPU0:P2#
```

Post-convergence, the RIB and LFIB also reflect the updated topology, showing both a primary path and a backup path calculated by TI-LFA, as shown in the following.

Post-convergence primary path in the RIB

After bringing up the link between P8 and P4, as shown in *Figure 9.2*, the ring is completed, resulting in a primary path from P2 to PE5 as well as a backup path to protect the primary path:

```
RP/0/RP0/CPU0:P2#show isis fast-reroute detail 5.5.5.5/32

L2 5.5.5.5/32 [50/115] Label: 16005, medium priority
    Installed Jan 01 07:32:55.873 for 00:01:49
        via 26.0.0.6, GigabitEthernet0/0/0/1, Label: 16005, P6, SRGB Base: 16000,
Weight: 0
            Backup path: LFA, via 23.0.0.3, GigabitEthernet0/0/0/0, Label: 16005,
P3, SRGB Base: 16000, Weight: 0, Metric: 120
        P: No, TM: 120, LC: No, NP: Yes, D: Yes, SRLG: Yes
    src PE5.00-00, 5.5.5.5, prefix-SID index 5, R:0 N:1 P:0 E:0 V:0 L:0,
Alg:0
RP/0/RP0/CPU0:P2#
```

The preceding output confirms that, while the primary path from P2 to PE5 is via router P6, there is now also a backup path via router P3.

This information is then downloaded into the FIB and the LFIB.

Post-convergence primary path in the LFIB

The following LFIB output confirms that the update in topology is reflected not only in the control plane but also in the data plane, where both the primary path and the backup path are installed:

```
RP/0/RP0/CPU0:P2#show mpls forwarding prefix 5.5.5.5/32 detail
Local  Outgoing    Prefix             Outgoing      Next Hop         Bytes
Label  Label       or ID              Interface                      Switched
------ ----------- ------------------ ------------- ---------------- ----------
--
16005  16005       SR Pfx (idx 5)     Gi0/0/0/1     26.0.0.6         0
       Updated: Jan  1 07:32:55.882
       Path Flags: 0x400 [  BKUP-IDX:0 (0xd674360) ]
       Version: 784, Priority: 1
       Label Stack (Top -> Bottom): { 16005 }
       NHID: 0x0, Encap-ID: N/A, Path idx: 1, Backup path idx: 0, Weight: 0
       MAC/Encaps: 4/8, MTU: 1500
       Outgoing Interface: GigabitEthernet0/0/0/1 (ifhandle 0x01000058)
       Packets Switched: 0

       16005       SR Pfx (idx
5)     Gi0/0/0/0   23.0.0.3          0             (!)
       Updated: Jan  1 07:32:55.882
       Path Flags: 0x300 [  IDX:0 BKUP, NoFwd ]
       Version: 784, Priority: 1
       Label Stack (Top -> Bottom): { 16005 }
       NHID: 0x0, Encap-ID: N/A, Path idx: 0, Backup path idx: 0, Weight: 0
       MAC/Encaps: 4/8, MTU: 1500
       Outgoing Interface: GigabitEthernet0/0/0/0 (ifhandle 0x01000020)
       Packets Switched: 0
       (!): FRR pure backup

  Traffic-Matrix Packets/Bytes Switched: 0/0
RP/0/RP0/CPU0:P2#
```

> **Note**
>
> The FIB output was skipped, but nonetheless, it would have reflected the same data from the RIB as seen earlier. Both the FIB and LFIB retrieve information from the RIB. The FIB output was omitted to concentrate on **Multiprotocol Label Switching (MPLS)**-to-MPLS labeled forwarding.

The emphasis of this lab was on configuring microloop avoidance exclusively on the P2 router. Nonetheless, in a production setting, it is advisable to implement microloop avoidance on all routers within the SR-MPLS domain.

The next section restores the topology to suit the requirements of the next lab.

Restore topology

The following configuration restores the topology to a certain state, setting the stage for the next chapter.

- On router P2, change the metric on the P2-P3 link back to 10 from 100:

P2

```
router isis IGP
interface GigabitEthernet0/0/0/0
  address-family ipv4 unicast
   metric 10
```

- On router P7, bring up the links toward P2, P3, and P4 back in IS-IS to activate them in the current topology:

P7

```
router isis IGP
interface GigabitEthernet0/0/0/1
  no shutdown
!
interface GigabitEthernet0/0/0/3
  no shutdown
!
interface GigabitEthernet0/0/0/4
  no shutdown
```

This concludes the microloop avoidance lab, and the lab summary follows.

Summary

In this chapter, the focus shifted from the previously discussed TI-LFA mechanism, which provides a backup path to protect traffic on the primary path, to exploring the microloop avoidance mechanism. This mechanism is crucial in scenarios where changes in network topology could cause traffic to briefly loop due to the delayed convergence of the IGP routing protocol, as routers update at different times.

The microloop avoidance mechanism works by creating an explicit MPLS tunnel for the primary path to the repair node, calculated using the same method as TI-LFA. This explicit tunnel, using MPLS labels and label stack, continues to forward traffic while IP routing converges.

On the Cisco IOS XR platform, a 60-second maximum time could be provisioned for network convergence. After this period, the explicit tunnel is replaced by the post-convergence path.

In the lab scenario, when the P8-P4 link was activated, traffic from P2 was expected to switch from the P2-P3 link to the P2-P6 link, as the cost to reach PE5 was lower through the P8-P4 link. However, since not all routers along the path may have converged, the microloop avoidance mechanism on router P2 triggered an explicit tunnel from P2 to P8. After the RIB update delay expired, the post-convergence path replaced the explicit tunnel.

Finally, the topology was restored to its original state in preparation for the next chapter. More TI-LFA scenarios will be covered in the next chapter.

References

The following is a list of references cited throughout this chapter, which provide additional details and insights on the topics discussed, including MPLS, **Fast Reroute** (**FRR**), and segment routing. These resources are recommended for further reading and deeper understanding.

- *[I-D.bashandy-rtgwg-segment-routing-uloop] Bashandy, A., Filsfils, C., Litkowski, S., Decraene, B., Francois, P., and P. Psenak, "Loop avoidance using Segment Routing", Work in Progress, Internet-Draft, draft- bashandy-rtgwg-segment-routing-uloop-16, 17 December 2023, <https://datatracker.ietf.org/doc/html/draft-bashandy-rtgwg-segment-routing-uloop-16>.*

- *[RFC7490] Bryant, S., Filsfils, C., Previdi, S., Shand, M., and N. So, "Remote Loop-Free Alternate (LFA) Fast Reroute (FRR)", RFC 7490, DOI 10.17487/RFC7490, April 2015, <https://www.rfc-editor.org/info/rfc7490>.*

- *[RFC7916] Litkowski, S., Ed., Decraene, B., Filsfils, C., Raza, K., Horneffer, M., and P. Sarkar, "Operational Management of Loop-Free Alternates", RFC 7916, DOI 10.17487/RFC7916, July 2016, <https://www.rfc-editor.org/info/rfc7916>.*

- *[RFC8029] Kompella, K., Swallow, G., Pignataro, C., Ed., Kumar, N., Aldrin, S., and M. Chen, "Detecting Multiprotocol Label Switched (MPLS) Data-Plane Failures", RFC 8029, DOI 10.17487/ RFC8029, March 2017, <https://www.rfc-editor.org/info/rfc8029>.*

- *[I-D.ietf-rtgwg-segment-routing-ti-lfa] Bashandy, A., Litkowski, S., Filsfils, C., Francois, P., Decraene, B., and D. Voyer, "Topology Independent Fast Reroute using Segment Routing", Work in Progress, Internet-Draft, draft-ietf-rtgwg-segment-routing-ti-lfa-13, 16 January 2024, <https://datatracker.ietf.org/doc/html/draft-ietf-rtgwg-segment-routing-ti-lfa-13>.*

- *[RFC5715] Shand, M. and S. Bryant, "A Framework for Loop-Free Convergence", RFC 5715, DOI 10.17487/RFC5715, January 2010, <https://www.rfc-editor.org/info/rfc5715>.*

- *[RFC6571] Filsfils, C., Ed., Francois, P., Ed., Shand, M., Decraene, B., Uttaro, J., Leymann, N., and M. Horneffer, "Loop-Free Alternate (LFA) Applicability in Service Provider (SP) Networks", RFC 6571, DOI 10.17487/RFC6571, June 2012, <https://www.rfc-editor.org/info/rfc6571>.*

10

Lab 9 – TI-LFA Node Protection

In the previous chapters, the primary focus was on TI-LFA's default protection method, which is link protection – a feature that remains active and cannot be disabled by default.

To further bolster network resilience, TI-LFA can be configured to incorporate additional constraints when determining backup paths. Among these, node protection and **Shared Risk Link Group** (SRLG) protection are widely utilized. These enhancements broaden protection capabilities by considering potential failures that extend beyond individual link failures.

This chapter explores the implementation of TI-LFA node protection specifically within an SR-MPLS network environment. Node protection ensures that if a node or its connecting link fails, traffic can efficiently reroute around the affected node, rather than solely relying on alternative paths for individual links.

Later in this book, a scenario will be presented in *Chapter 13* where TI-LFA cannot provide a backup path using two additional labels, unlike in the double-segment FRR scenario. This limitation arises due to additional constraints such as node or SRLG protection.

In such situations, the IGP (in this case, IS-IS) using TI-LFA initiates a **Segment Routing Auto-Tunnel** (**SRAT**) as the backup path. SRAT is similar to **Segment Routing Traffic Engineering** (**SR-TE**), but SRTE's features and applications extend far beyond auto-tunnel instantiation, including policy-based routing. In earlier IOS XR versions, SRAT was not mentioned in the output, and SR-TE is also outside the scope of this book. The key concept here is that IGP creates a TI-LFA auto-tunnel backup path, and to avoid confusion with policy-based SR-TE tunnels, this is now referred to as SRAT.

For SRAT to function effectively on the Cisco IOS XR platform, MPLS traffic engineering must be enabled. This configuration ensures that the auto-tunnel is properly injected into the **Routing Information Base** (**RIB**), **Forwarding Information Base** (**FIB**), and **Label Forwarding Information Base** (**LFIB**).

Objective

The objective of this lab is to explore a scenario where TI-LFA is instructed to provide node protection, meaning the PLR calculates and installs a fast reroute path for a potential link failure to circumvent the entire router, not just the link. This is accomplished using the following tasks:

1.   Applying the configuration to enable TI-LFA node protection on P2

2.   Verifying that the TI-LFA node protection backup path is computed and added to the RIB, FIB, and LFIB of P2.

Therefore, the next step involves configuring all routers in the network with MPLS traffic engineering to facilitate the deployment and operation of SRAT.

> **Note**
> For the purpose of this book, MPLS traffic engineering is not explored beyond the commands necessary to support auto-tunnels.

Let's dive in!

MPLS TE (MPLS Traffic Engineering)

As previously discussed in the book, the maximum number of labels supported in the label stack of the backup path is 3. In situations where more than three labels are needed, a traffic-engineering tunnel is initiated, requiring the following configuration on all routers in the topology to source the tunnel from the loopback IP address of the router.

Enabling MPLS TE on all routers of the topology

The following configuration is required to enable MPLS TE for the IS-IS link state routing protocol on the routers. This allows a router to create TI-LFA backup paths using MPLS TE tunnels.

Let's apply the following configuration on all the routers to enable MPLS TE on them:

```
ipv4 unnumbered mpls traffic-eng Loopback0

mpls traffic-eng

router isis IGP
  address-family ipv4 unicast
  mpls traffic-eng router-id Loopback0
!
```

In summary, this configuration snippet accomplishes the following:

- It enables MPLS TE tunnel interfaces to use the IP address of Loopback0, as they do not have their own IP address.

- It globally enables MPLS TE on the router.

- It configures IS-IS to support MPLS TE for IPv4 unicast routing.

Next, we will prepare the topology for the upcoming lab scenarios.

Preparing the base topology

Implement the provided configuration to establish the desired topology shown in *Figure 10.1*. This topology will be used to explore TI-LFA node protection and TI-LFA SRLG protection (in the next chapter and onwards):

- Increase the metric on the P6 router for the link connecting to the P7 router from 10 to 100:

  P6

  ```
  router isis IGP
  interface GigabitEthernet0/0/0/0
    address-family ipv4 unicast
     metric 100
  ```

- Increase the metric on the P7 router for the link connecting to the P6 and P4 routers from 10 to 100:

  P7

  ```
  router isis IGP
  interface GigabitEthernet0/0/0/0
    address-family ipv4 unicast
     metric 100
     !
   !
  interface GigabitEthernet0/0/0/4
    address-family ipv4 unicast
     metric 100
  ```

- Increase the metric on the P4 router for the link connecting to the P7 router from 10 to 100:

  P4

  ```
  router isis IGP
  interface GigabitEthernet0/0/0/4
    address-family ipv4 unicast
     metric 100
  ```

- The P8 router will not be included in the topology to explore node protection and SRLG-protection scenarios. Hence, shut down its connections to P7 and P4 to isolate it from the network, as shown here:

P8

```
router isis IGP
interface GigabitEthernet0/0/0/1
  shutdown
!
interface GigabitEthernet0/0/0/2
  shutdown
```

The preceding configuration prepares the following diagram of the topology.

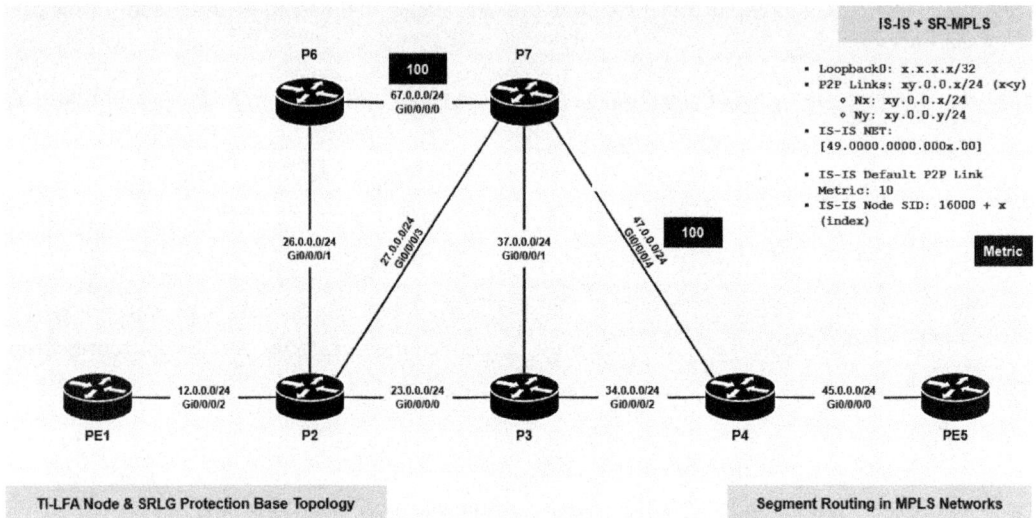

Figure 10.1 – The TI-LFA node and SRLG protection base topology

The lab scenario will focus on the P2 router to protect traffic toward the destination, PE5.

Note

For the upcoming labs and chapters, TI-LFA protection mechanisms will be enabled only on the P2 router, the **Point of Local Repair** (**PLR**), to protect traffic from PE1 to PE5. The P2 router and its connections in the topology are strategically designed for this purpose, allowing all protection scenarios to be explored on this router exclusively.

In production networks, it is recommended to enable protection on all routers within the SR-MPLS domain to ensure that traffic is protected throughout the entire network, in both forward and reverse directions.

Configuration

To enable node protection, configure the following tiebreaker settings with a specified tiebreaker index. The significance of the index will be covered later in *Chapter 14* of the book.

The TI-LFA node protection in the context of router IS-IS is enabled, as shown in the following snippet. A tiebreaker index must be configured, and the value 200 has been chosen strategically for this book. Its significance will be examined during the tiebreaker scenario discussed later:

P2

```
router isis IGP
address-family ipv4 unicast
  fast-reroute per-prefix tiebreaker node-protecting index 200
```

After activation, the TI-LFA backup path calculation on the P2 router includes node protection as an additional constraint for all primary paths through the P2–P3 link, and it avoids transiting through the P3 router, as shown in the following diagram.

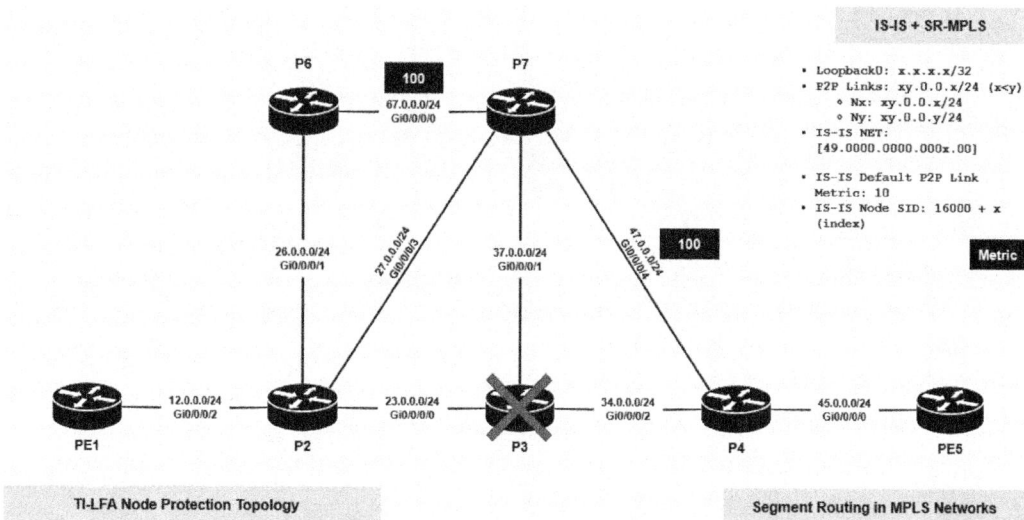

Figure 10.2 – TI-LFA node protection

The subsequent section validates the backup path.

> **Note**
>
> The output of the show isis interface command on P2 will display an index value of 200 in the Node Protecting column within the TI-LFA section (as shown in *Chapter 5*), indicating that node protection is now enabled.

Verification

The following output indicates that the node protection mechanism actively calculates backup paths to bypass the entire router, not just the link.

Backup path in the RIB

On the P2 router, if there was a failure in the P2–P3 link or a failure of the P3 node, the backup path would circumvent the P3 node entirely.

The P7 router has a more optimal path to PE5 through P3, but it hasn't been used due to the presence of a node protection request.

Instead, P7 is instructed to utilize the adjacency toward P4 to meet the node protection request:

```
RP/0/RP0/CPU0:P2#show isis fast-reroute detail 5.5.5.5/32


L2 5.5.5.5/32 [30/115] Label: 16005, medium priority
    Installed Jan 01 07:39:23.614 for 00:00:06
       via 23.0.0.3, GigabitEthernet0/0/0/0, Label: 16005, P3, SRGB Base: 16000,
Weight: 0
          Backup path: TI-LFA (node), via 27.0.0.7, GigabitEthernet0/0/0/3 P7,
SRGB Base: 16000, Weight: 0, Metric: 120
            P node: P7.00 [7.7.7.7], Label: ImpNull
            Q node: P4.00 [4.4.4.4], Label: 24023
            Prefix label: 16005
            Backup-src: PE5.00
          P: No, TM: 120, LC: No, NP: Yes, D: No, SRLG: Yes
       src PE5.00-00, 5.5.5.5, prefix-SID index 5, R:0 N:1 P:0 E:0 V:0 L:0,
Alg:0
RP/0/RP0/CPU0:P2#
```

This validates the proper functioning of node protection on P2.

> **Note**
> The preceding output indicates `TI-LFA (node)`, which essentially means `TI-LFA (node+link)`, since link protection is enabled by default and cannot be disabled.

```
RP/0/RP0/CPU0:P2#show route 5.5.5.5/32 detail

Routing entry for 5.5.5.5/32
  Known via "isis IGP", distance 115, metric 30, labeled SR, type level-2
  Installed Jan  1 07:39:23.614 for 00:00:26
  Routing Descriptor Blocks
    23.0.0.3, from 5.5.5.5, via GigabitEthernet0/0/0/0, Protected
      Route metric is 30
```

```
      Label: 0x3e85 (16005)
      Tunnel ID: None
      Binding Label: None
      Extended communities count: 0
      Path id:1        Path ref count:0
      NHID:0x7(Ref:12)
      Backup path id:65
    27.0.0.7, from 5.5.5.5, via GigabitEthernet0/0/0/3, Backup (TI-LFA)
      Repair Node(s): 7.7.7.7, 4.4.4.4
      Route metric is 120
      Labels: 0x3 0x5dd7 0x3e85 (3 24023 16005)
      Tunnel ID: None
      Binding Label: None
      Extended communities count: 0
      Path id:65              Path ref count:1
      NHID:0x9(Ref:14)
  Route version is 0x46 (70)
  Local Label: 0x3e85 (16005)
  IP Precedence: Not Set
  QoS Group ID: Not Set
  Flow-tag: Not Set
  Fwd-class: Not Set
  Route Priority: RIB_PRIORITY_NON_RECURSIVE_MEDIUM (7) SVD Type RIB_SVD_TYPE_
LOCAL
  Download Priority 1, Download Version 961
  No advertising protos.
RP/0/RP0/CPU0:P2#
```

The presence of `ImpNull` and label 3 in the label stack output indicates the implicit null label. The P7 router is adjacent to the P2 router; therefore, **Penultimate Hop Popping** (PHP) occurs.

Verify that the LFIB has downloaded the same information from the RIB, as described in the next section.

Backup path in the LFIB

The provided output confirms that the backup route passes through P7 in the LFIB:

```
RP/0/RP0/CPU0:P2#show mpls forwarding prefix 5.5.5.5/32
Local   Outgoing    Prefix              Outgoing      Next Hop         Bytes
Label   Label       or ID               Interface                      Switched
------  ----------- ------------------- ------------- ---------------- ----------
--
16005   16005       SR Pfx (idx 5)      Gi0/0/0/0     23.0.0.3         0
        Pop         SR Pfx (idx
5)      Gi0/0/0/3   27.0.0.7            0             (!)
RP/0/RP0/CPU0:P2#
```

Upon examining the detailed output, it is revealed that the label stack is identical to the one observed in the RIB. This configuration circumvents the P3 node entirely, providing node protection:

```
RP/0/RP0/CPU0:P2#show mpls forwarding prefix 5.5.5.5/32 detail
Local  Outgoing    Prefix           Outgoing      Next Hop         Bytes
Label  Label       or ID            Interface                      Switched
------ ----------- ---------------- ------------- ---------------- ----------
--
16005  16005       SR Pfx (idx 5)   Gi0/0/0/0     23.0.0.3         0
       Updated: Jan  1 07:39:23.621
       Path Flags: 0x400 [  BKUP-IDX:1 (0xd674810) ]
       Version: 961, Priority: 1
       Label Stack (Top -> Bottom): { 16005 }
       NHID: 0x0, Encap-ID: N/A, Path idx: 0, Backup path idx: 1, Weight: 0
       MAC/Encaps: 4/8, MTU: 1500
       Outgoing Interface: GigabitEthernet0/0/0/0 (ifhandle 0x01000020)
       Packets Switched: 0

       Pop         SR Pfx (idx
5)     Gi0/0/0/3   27.0.0.7         0             (!)
       Updated: Jan  1 07:39:23.621
       Path Flags: 0xb00 [  IDX:1 BKUP, NoFwd ]
       Version: 961, Priority: 1
       Label Stack (Top -> Bottom): { Imp-Null 24023 16005 }
       NHID: 0x0, Encap-ID: N/A, Path idx: 1, Backup path idx: 0, Weight: 0
       MAC/Encaps: 4/12, MTU: 1500
       Outgoing Interface: GigabitEthernet0/0/0/3 (ifhandle 0x01000048)
       Packets Switched: 0
       (!): FRR pure backup

    Traffic-Matrix Packets/Bytes Switched: 0/0
RP/0/RP0/CPU0:P2#
```

This confirms that the TI-LFA node protection mechanism functions as expected on the P2 router. The backup path avoids the P3 router to ensure node protection.

In the following section, the backup path traceroute is verified.

Verifying a backup path traceroute

The essential components necessary for the nil-FEC traceroute, as derived from the RIB, FIB, or LFIB, are as follows:

- **Label stack (top to bottom)**: 3, 24023, 16005, directing traffic toward P7, then P7– P4, and finally, PE5

- **Outgoing interface**: GigabitEthernet0/0/0/3, toward P7.

- **Next-hop IP address of the outgoing interface**: 27.0.0.7, toward P7.

The following output traces the backup path through all the hops, as per the requested label stack, outgoing interface, and the next hop, which was computed by the TI-LFA in the previous output:

```
RP/0/RP0/CPU0:P2#traceroute sr-mpls nil-fec labels 3,24023,16005 output
interface GigabitEthernet0/0/0/3 nexthop 27.0.0.7


Tracing MPLS Label Switched Path with Nil FEC with labels [3,24023,16005],
timeout is 2 seconds

Codes: '!' - success, 'Q' - request not sent, '.' - timeout,
  'L' - labeled output interface, 'B' - unlabeled output interface,
  'D' - DS Map mismatch, 'F' - no FEC mapping, 'f' - FEC mismatch,
  'M' - malformed request, 'm' - unsupported tlvs, 'N' - no rx label,
  'P' - no rx intf label prot, 'p' - premature termination of LSP,
  'R' - transit router, 'I' - unknown upstream index,
  'X' - unknown return code, 'x' - return code 0


Type escape sequence to abort.

  0 27.0.0.2 MRU 1500 [Labels: implicit-null/24023/16005/explicit-null Exp:
0/0/0/0]
L 1 27.0.0.7 MRU 1500 [Labels: implicit-null/16005/explicit-null Exp: 0/0/0]
13 ms
L 2 47.0.0.4 MRU 1500 [Labels: implicit-null/explicit-null Exp: 0/0] 13 ms
! 3 45.0.0.5 12 ms
RP/0/RP0/CPU0:P2#
```

The successful completion of the trace confirms that if there is a P2–P3 link failure, the fast reroute traffic on the P2 router, following the aforementioned label stack, outgoing interface, and next hop, would bypass the P3 router entirely and be forwarded to the destination router, PE5, ensuring node protection.

The subsequent section elaborates on how TI-LFA incorporates the additional requirement into the backup path calculation.

Understanding the backup path calculation

Figure 10.3 illustrates the topology diagram for TI-LFA node protection.

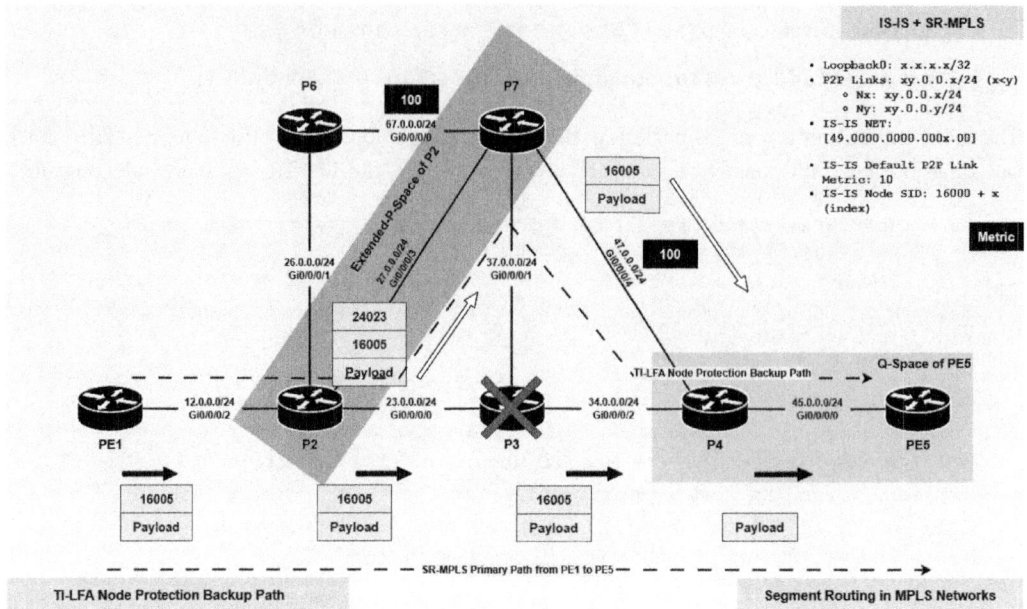

Figure 10.3 – The TI-LFA node protection primary and backup paths

The following explanation details the computation and selection of the primary path and backup path.

Primary path

The primary path from P2 to PE5 goes via P3, with the end-to-end path, cost of 30 (P2 - P3 - P4 - PE5), making it the shortest path.

Backup path

Moreover, the backup path is classified as **TI-LFA (node)** because node protection is explicitly requested in the P2 router configuration.

TI-LFA

In this scenario, the post-convergence path would be P2 - P7 - P4 - PE5 with a cost of 120.

In addressing the node failure on P3, TI-LFA provides node protection by collecting the following information:

- **P-Space of P2 along the post-convergence path**: The group of routers on the post-convergence path, reachable from P2 through the shortest path tree, excluding the traversal of the P3 node, forms the P-Space of P2.

    In this topology, these routers are P7.

- **Extended P-Space of P2 along the post-convergence path**: The extended P-Space of P2 is a combination of P2's P-Space and the P-Space of P2's neighbors along the post-convergence path. This can be determined by calculating the shortest path tree, rooted at each of P2's neighbors, and removing the subtree accessed via the P3 node (including the ECMP nodes).

  There are no routers in the P-Space of P7.

  Therefore, the extended P-Space of P2 along the post-convergence path includes only P7.

- **Q-Space of PE5 along the post-convergence path**: The group of routers along the post-convergence path, through which the PE5 router can be reached via the shortest path tree without traversing the P3 node, forms the Q-Space of PE5.

  The Q-Space of PE5 comprises P4.

The TI-LFA repair path is a sequence of segments (a stack of labels) along the post-convergence path on the outgoing interface. This path is determined by the intersection of the extended P-Space of P2 and the Q-Space of PE5.

There is no overlap between the extended P-Space and Q-Space in this scenario. The deeper downstream P node is P7, and the adjacent Q node to it is P4.

The endpoint for the repair tunnel should be P4 along the post-convergence path. To accomplish this, P2 encodes additional information in the label stack, instructing P7 to use the adjacency toward P4, and provides additional instructions to ensure a loop-free alternate.

This is evident in the output, where the outgoing label on the backup path from the PLR, P2, to the destination, PE5, is the Node-SID of P7 and the Adjacency-SID on P7 to P4.

If there is a P2–P3 link failure or a P3 node failure, the P2 router redirects the traffic to the backup path, leading it through P7 and then P4. Thus, the TI-LFA backup path ensures loop-free forwarding with node protection.

Restoring the topology

Let's now restore the topology for the benefit of the next chapter. Remove node protection from the P2 router, as shown here:

P2

```
router isis IGP
address-family ipv4 unicast
  no fast-reroute per-prefix tiebreaker node-protecting index 200
```

This concludes the TI-LFA node protection scenario.

Summary

The chapter began with an introduction to MPLS TE within the network topology. This introduction is intended to facilitate the creation of auto-tunnels, which will be discussed later in the book. Introducing MPLS TE at this point provides a foundation before delving further into the TI-LFA lab scenarios.

In scenarios where the TI-LFA backup path could not be identified with two additional segments encoded as labels – given that the maximum label depth is three for the Cisco IOS XR platform – MPLS TE initiates an auto-tunnel instead.

The topology was then modified to explore node protection and SRLG protection scenarios for ease of understanding.

Node protection is enabled by configuring a non-zero fast-reroute tiebreaker index value on P2. The significance of this index will be discussed later in the book.

The TI-LFA backup path outputs from the RIB, FIB, and LFIB confirmed that the desired node protection backup path was available. Additionally, the backup path MPLS traceroute validated this configuration.

The TI-LFA backup path calculation demonstrated that the entire P3 node, through which the next hop of the primary path was traversing, was omitted from the post-convergence path calculation. Subsequently, the extended P-Space and Q-Space nodes were identified to create the repair tunnel.

Finally, node protection was removed, which is necessary before we proceed to the next chapter to explore SRLG protection.

References

The following references provide additional details and sources cited throughout the chapter. They offer further reading and background on key concepts such as **Fast Reroute (FRR)**, **Loop-Free Alternates (LFA)**, and Segment Routing in MPLS networks:

- [RFC7490] Bryant, S., Filsfils, C., Previdi, S., Shand, M., and N. So, "Remote Loop-Free Alternate (LFA) Fast Reroute (FRR)", RFC 7490, DOI 10.17487/RFC7490, April 2015, <https://www.rfc-editor.org/info/rfc7490>.

- [RFC7916] Litkowski, S., Ed., Decraene, B., Filsfils, C., Raza, K., Horneffer, M., and P. Sarkar, "Operational Management of Loop-Free Alternates", RFC 7916, DOI 10.17487/RFC7916, July 2016, <https://www.rfc-editor.org/info/rfc7916>.

- [RFC8029] Kompella, K., Swallow, G., Pignataro, C., Ed., Kumar, N., Aldrin, S., and M. Chen, "Detecting Multiprotocol Label Switched (MPLS) Data-Plane Failures", RFC 8029, DOI 10.17487/RFC8029, March 2017, <https://www.rfc-editor.org/info/rfc8029>.

- [I-D.ietf-rtgwg-segment-routing-ti-lfa] Bashandy, A., Litkowski, S., Filsfils, C., Francois, P., Decraene, B., and D. Voyer, "Topology Independent Fast Reroute using Segment Routing", Work in Progress, Internet-Draft, draft-ietf-rtgwg-segment-routing-ti-lfa-13, 16 January 2024, <https://datatracker.ietf.org/doc/html/draft-ietf-rtgwg-segment-routing-ti-lfa-13>.

- [RFC5715] Shand, M. and S. Bryant, "A Framework for Loop-Free Convergence", RFC 5715, DOI 10.17487/RFC5715, January 2010, <https://www.rfc-editor.org/info/rfc5715>.

- [RFC6571] Filsfils, C., Ed., Francois, P., Ed., Shand, M., Decraene, B., Uttaro, J., Leymann, N., and M. Horneffer, "Loop-Free Alternate (LFA) Applicability in Service Provider (SP) Networks", RFC 6571, DOI 10.17487/RFC6571, June 2012, <https://www.rfc-editor.org/info/rfc6571>.

Lab 10 – TI-LFA Local SRLG-Disjoint Protection

A **Shared Risk Link Group** (**SRLG**) denotes a set of links that share a common risk, potentially being physically routed along the same path with common transmission. This concept holds significance in the design and analysis of network resilience, as a failure within an SRLG can result in correlated impacts. If one link in the SRLG fails, it might lead to the failure of others in the group. In such a scenario, if the backup path of one link is routed along the primary path within the same SRLG, the backup would also fail when the primary path is affected.

In such cases, it is vital to guide the backup path away from SRLG links, ensuring SRLG protection.

The TI-LFA SRLG protection method in SR-MPLS networks can create a backup path upon request, mitigating potential issues associated with SRLG.

In the previous chapter, P2 requested node protection, and TI-LFA provided a post-convergence backup path by entirely omitting the next-hop router, P3, routing instead through P7 and P4.

The SRLG protection scenario depicted in *Figure 11.1* illustrates that router P2 is initially informed about local links sharing the same transmission medium, specifically, P2-P3 and P2-P7. This implies that if the P2-P3 link fails, the P2-P7 link will also fail.

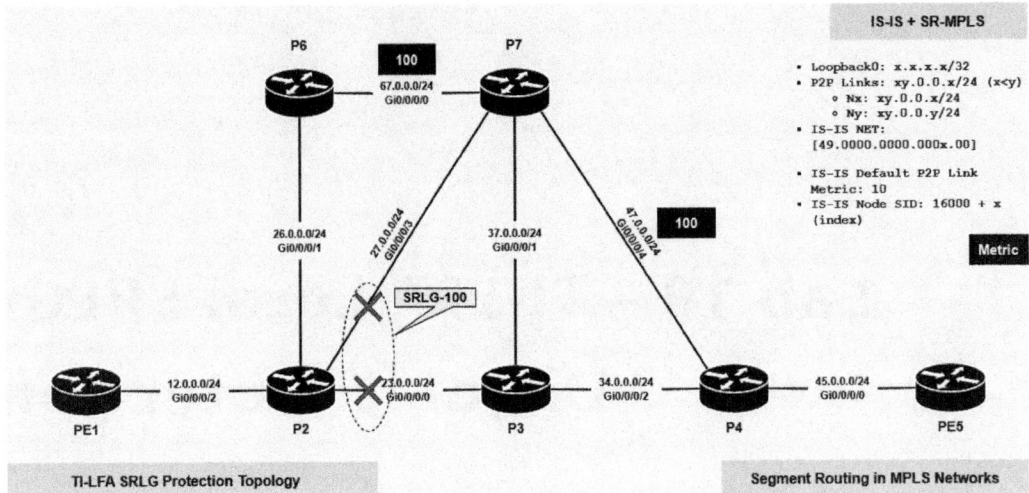

Figure 11.1 – TI-LFA SRLG protection

If a fast reroute backup path is protecting the P2-P3 link by rerouting it over the P2-P7 link, and the P2-P3 link fails, the backup path would also fail, potentially causing traffic to be blackholed until IGP convergence completes and establishes a new primary path.

The TI-LFA SRLG protection method alleviates this issue by ensuring that the backup path also avoids links within the SRLG.

Objective

The objective of this lab is to explore a scenario where TI-LFA is configured to provide SRLG protection. This involves instructing the **Point of Local Repair (PLR)** router to calculate and install a fast reroute path to be used in the event of a link failure, aiming to avoid not only the failed link but also others that share the same risk of failure. The local SRLG-disjoint protection is accomplished using the following tasks:

1. Inform router P2 about its local SRLGs.

2. Apply the configuration to enable TI-LFA SRLG protection on P2.

3. Verify that the local SRLG protection backup path is computed and added to the RIB, FIB, and LFIB of P2.

The task begins by first identifying the shared risks in the network with the help of the transmission team. Once this is done, each SRLG is allocated a unique identity and configured on the routers.

The SRLGs in this case, as explored in this lab, are local SRLGs. This means they are only locally significant to the PLR router, P2, and remote routers will not be aware of them, and vice versa.

The next section covers the identification and configuration of shared risks in the network.

Configuration

We will now configure the interfaces to belong to the same SRLG and instruct TI-LFA to provide SRLG protection in the next section.

The following configuration sets up an SRLG group named SRLG-100 on the Gi0/0/0/0 and Gi0/0/0/1 interfaces on router P2, as shown in *Figure 11.1*, recognizing them as sharing the same risk:

P2

```
srlg
name SRLG-100 value 100
!
interface GigabitEthernet0/0/0/0
  name SRLG-100
!
interface GigabitEthernet0/0/0/3
  name SRLG-100
!
router isis IGP
address-family ipv4 unicast
  fast-reroute per-prefix tiebreaker srlg-disjoint index 100
```

Although the SRLG group name is locally significant to router P2, the value of 100 can later hold significance network-wide, as further explored in the Global Weighted SRLG chapter.

SRLG protection is then requested as part of the tiebreaker configuration under the IS-IS routing protocol, with an index value of 100, as shown next.

SRLG-100 and the tiebreaker index value of 100 are unrelated; their matching numbers are purely coincidental.

The significance of choosing the index value will be explored later in *Chapter 14*.

With local SRLG-disjoint protection now enabled, the next section covers the verification process.

Verification

Upon checking the RIB, the output clearly indicates that in the event of a P2-P3 link failure, the backup path from P2 to the destination router PE5 avoids using the P2-P7 link. This behavior aligns with the configuration within the same SRLG, where SRLG protection is requested.

Backup path in the RIB

The backup path now traverses via P6 and then P7 instead:

```
RP/0/RP0/CPU0:P2#show isis fast-reroute detail 5.5.5.5/32

L2 5.5.5.5/32 [30/115] Label: 16005, medium priority
    Installed Jan 01 07:43:32.347 for 00:00:14
      via 23.0.0.3, GigabitEthernet0/0/0/0, Label: 16005, P3, SRGB Base: 16000,
Weight: 0
         Backup path: TI-LFA (srlg), via 26.0.0.6, GigabitEthernet0/0/0/1 P6,
SRGB Base: 16000, Weight: 0, Metric: 140
         P node: P6.00 [6.6.6.6], Label: ImpNull
         Q node: P7.00 [7.7.7.7], Label: 24023
         Prefix label: 16005
         Backup-src: PE5.00
       P: No, TM: 140, LC: No, NP: No, D: No, SRLG: Yes
     src PE5.00-00, 5.5.5.5, prefix-SID index 5, R:0 N:1 P:0 E:0 V:0 L:0,
Alg:0
RP/0/RP0/CPU0:P2#
```

> **Note**
>
> The preceding output indicates TI-LFA (srlg), which essentially means TI-LFA
> (srlg+link) since link protection is enabled by default and cannot be disabled.

The following routing table output confirms that routers P6 and P7 are the repair nodes providing local SRLG-disjoint protection:

```
RP/0/RP0/CPU0:P2#show route 5.5.5.5/32

Routing entry for 5.5.5.5/32
   Known via "isis IGP", distance 115, metric 30, labeled SR, type level-2
   Installed Jan  1 07:43:32.347 for 00:00:43
   Routing Descriptor Blocks
     23.0.0.3, from 5.5.5.5, via GigabitEthernet0/0/0/0, Protected
       Route metric is 30
     26.0.0.6, from 5.5.5.5, via GigabitEthernet0/0/0/1, Backup (TI-LFA)
       Repair Node(s): 6.6.6.6, 7.7.7.7
       Route metric is 140
   No advertising protos.
RP/0/RP0/CPU0:P2#
```

The FIB and LFIB program the forwarding plane according to the TI-LFA backup path calculation in the RIB, as shown next.

Backup path in the FIB

The backup path details and the label stack correspond to the output observed in the RIB:

```
RP/0/RP0/CPU0:P2#show cef 5.5.5.5/32 brief
5.5.5.5/32, version 995, labeled SR, internal 0x1000001 0x8310 (ptr 0xe7ac838)
[1], 0x600 (0xda5f070), 0xa28 (0x2218f758)
Updated Jan  1 07:43:32.365
remote adjacency to GigabitEthernet0/0/0/0
Prefix Len 32, traffic index 0, precedence n/a, priority 1
   via 23.0.0.3/32, GigabitEthernet0/0/0/0, 14 dependencies, weight 0, class
0, protected [flags 0x400]
    path-idx 0 bkup-idx 1 NHID 0x0 [0xd674630 0xd674630]
    next hop 23.0.0.3/32
     local label 16005      labels imposed {16005}
   via 26.0.0.6/32, GigabitEthernet0/0/0/1, 18 dependencies, weight 0, class
0, backup (TI-LFA) [flags 0xb00]
    path-idx 1 NHID 0x0 [0x22032470 0x0]
    next hop 26.0.0.6/32, Repair Node(s): 6.6.6.6, 7.7.7.7
    remote adjacency
     local label 16005      labels imposed {ImplNull 24023 16005}
RP/0/RP0/CPU0:P2#
```

The TI-LFA local SRLG-disjoint protection is provided in the FIB as requested.

Backup path in the LFIB

The same information is downloaded in the LFIB from the RIB:

```
RP/0/RP0/CPU0:P2#show mpls forwarding prefix 5.5.5.5/32
Local  Outgoing   Prefix            Outgoing     Next Hop        Bytes
Label  Label      or ID             Interface                    Switched
------ ---------- ----------------- ------------ --------------- ----------
--
16005  16005      SR Pfx (idx 5)    Gi0/0/0/0    23.0.0.3        0
       Pop        SR Pfx (idx
5)     Gi0/0/0/1  26.0.0.6          0            (!)
RP/0/RP0/CPU0:P2#
```

The following detailed output reveals the same label stack as in the RIB:

```
RP/0/RP0/CPU0:P2#show mpls forwarding prefix 5.5.5.5/32 detail
Local  Outgoing   Prefix            Outgoing     Next Hop        Bytes
Label  Label      or ID             Interface                    Switched
------ ---------- ----------------- ------------ --------------- ----------
--
16005  16005      SR Pfx (idx 5)    Gi0/0/0/0    23.0.0.3        0
     Updated: Jan  1 07:43:32.372
```

```
      Path Flags: 0x400 [ BKUP-IDX:1 (0xd674630) ]
      Version: 995, Priority: 1
      Label Stack (Top -> Bottom): { 16005 }
      NHID: 0x0, Encap-ID: N/A, Path idx: 0, Backup path idx: 1, Weight: 0
      MAC/Encaps: 4/8, MTU: 1500
      Outgoing Interface: GigabitEthernet0/0/0/0 (ifhandle 0x01000020)
      Packets Switched: 0

      Pop         SR Pfx (idx
5)    Gi0/0/0/1   26.0.0.6        0                   (!)
      Updated: Jan  1 07:43:32.372
      Path Flags: 0xb00 [ IDX:1 BKUP, NoFwd ]
      Version: 995, Priority: 1
      Label Stack (Top -> Bottom): { Imp-Null 24023 16005 }
      NHID: 0x0, Encap-ID: N/A, Path idx: 1, Backup path idx: 0, Weight: 0
      MAC/Encaps: 4/12, MTU: 1500
      Outgoing Interface: GigabitEthernet0/0/0/1 (ifhandle 0x01000058)
      Packets Switched: 0
      (!): FRR pure backup

   Traffic-Matrix Packets/Bytes Switched: 0/0
RP/0/RP0/CPU0:P2#
```

The TI-LFA local SRLG-disjoint protection is provided in the LFIB as requested.

The next section will validate the backup path in the data plane.

Verifying the backup path traceroute

The essential components necessary for the nil-fec traceroute, as derived from the RIB, FIB, or LFIB, are as follows:

- **Label stack (top to bottom)**: 3, 24023, 16005, directing traffic to P6, then P6-P7, and finally to PE5

- **Outgoing interface**: GigabitEthernet0/0/0/1, toward P6

- **Next-hop IP address of outgoing interface**: 26.0.0.6, to P6

The following output traces the backup path through all the hops, as per the requested label stack, outgoing interface, and the next hop, which was computed by the TI-LFA in the previous output:

```
RP/0/RP0/CPU0:P2#traceroute sr-mpls nil-fec labels 3,24023,16005 output
interface GigabitEthernet0/0/0/1 nexthop 26.0.0.6

Tracing MPLS Label Switched Path with Nil FEC with labels [3,24023,16005],
timeout is 2 seconds

Codes: '!' - success, 'Q' - request not sent, '.' - timeout,
  'L' - labeled output interface, 'B' - unlabeled output interface,
  'D' - DS Map mismatch, 'F' - no FEC mapping, 'f' - FEC mismatch,
  'M' - malformed request, 'm' - unsupported tlvs, 'N' - no rx label,
  'P' - no rx intf label prot, 'p' - premature termination of LSP,
  'R' - transit router, 'I' - unknown upstream index,
  'X' - unknown return code, 'x' - return code 0

Type escape sequence to abort.

  0 26.0.0.2 MRU 1500 [Labels: implicit-null/24023/16005/explicit-null Exp:
0/0/0/0]
L 1 26.0.0.6 MRU 1500 [Labels: implicit-null/16005/explicit-null Exp: 0/0/0]
13 ms
L 2 67.0.0.7 MRU 1500 [Labels: 16005/explicit-null Exp: 0/0] 10 ms
L 3 37.0.0.3 MRU 1500 [Labels: 16005/explicit-null Exp: 0/0] 12 ms
L 4 34.0.0.4 MRU 1500 [Labels: implicit-null/explicit-null Exp: 0/0] 20 ms
! 5 45.0.0.5 14 ms
RP/0/RP0/CPU0:P2#
```

The successful traceroute confirms that in the event of a P2 - P3 link failure on router P2, the TI-LFA path uses the aforementioned label stack, outgoing interface, and next hop to successfully fast-reroute traffic to the destination, ensuring local SRLG-disjoint protection.

The next section elaborates on how TI-LFA incorporated the additional requirement into the backup path calculation.

Understanding the backup path calculation

The following topology diagram describes the backup path calculation:

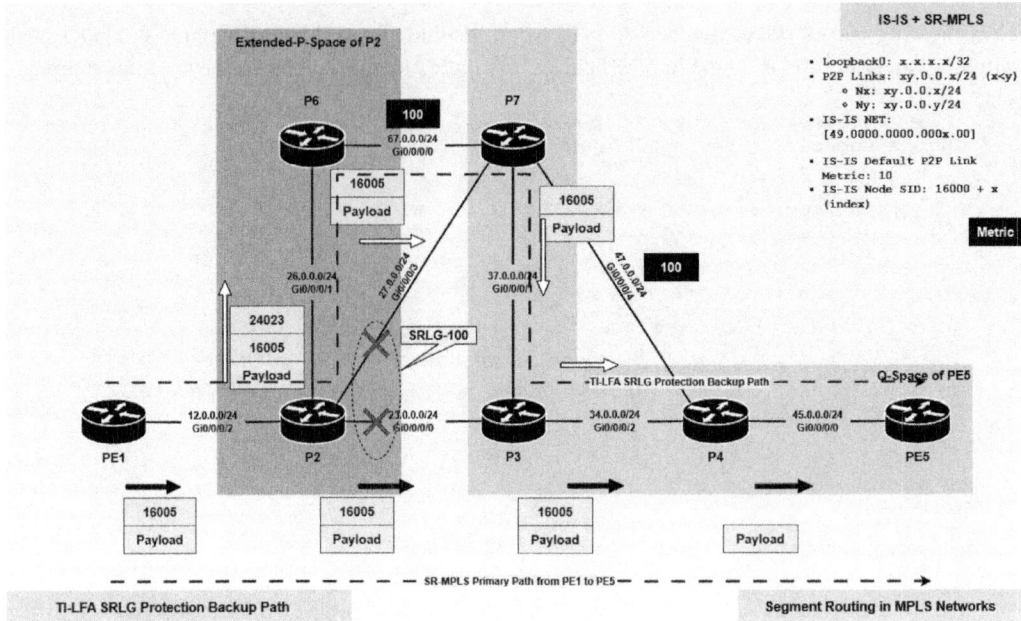

Figure 11.2 – TI-LFA SRLG protection primary and backup paths

The detailed explanation of the computation and selection of the primary path and backup path for the preceding topology is as follows.

Primary path

The primary path from P2 to PE5 goes via P3, with the end-to-end path cost of 30 (P2 - P3 - P4 - PE5), making it the shortest path.

Backup path

The backup path is classified as TI-LFA (srlg) because SRLG protection is explicitly requested in the P2 router configuration:

TI-LFA

In addressing the link failure on the P2-P3 link, TI-LFA provides SRLG protection by collecting the following information:

The post-convergence path in the scenario, including the local srlg-disjoint request, would be P2 - P6 - P7 - P3 - P4 - PE5, with a cost of 140.

- **P-Space of P2 along the post-convergence path**: The group of routers on the post-convergence path reachable from P2 through the shortest path tree, excluding the traversal of the P2-P3 link and P2-P7 link, forms the P-Space of P2.

In this topology, that router is P6.

- **Extended P-Space of P2 along the post-convergence path**: The extended P-Space of P2 is the combination of P2's P-Space and the P-Space of P2's neighbors along the post-convergence path. This can be determined by calculating the shortest path tree rooted at each of P2's neighbors and removing the subtree accessed via the P2 - P3 link and P2 - P7 link (including the ECMP nodes).

  There are no routers in the P-Space of P6.

  Therefore, the extended P-Space of P2 along the post-convergence path includes only P6.

- **Q-Space of PE5 along the post-convergence path**: The group of routers along the post-convergence path, through which router PE5 can be reached via the shortest path tree without traversing the P2 - P3 link and P2 - P7 link, forms the Q-Space of PE5.

  The Q-Space of PE5 comprises P7, P3, and P4.

The TI-LFA repair path is a sequence of segments (a stack of labels) along the post-convergence path on the outgoing interface. This path is determined by the intersection of the extended P-Space of P2 and the Q-Space of PE5.

There is no overlap between the extended P-Space and Q-Space in this scenario. The deeper downstream P node is P6, and the adjacent Q node to it is P7.

The endpoint for the repair tunnel should be P7 along the post-convergence path. To accomplish this, P2 encodes additional information in the label stack, instructing P6 to use the adjacency toward P7 and providing additional instructions to ensure a loop-free alternate.

This is evident in the output, where the outgoing label on the backup path from the PLR P2 to the destination PE5 is the `implicit-null` label toward P6, followed by the Adjacency-SID label of the P6 - P7 link.

In the event of a P2 - P3 link failure, router P2 redirects the traffic to the backup path, routing it through P6 and then via P6 - P7 to P7.

Thus, the TI-LFA backup path ensures loop-free forwarding with local-SRLG disjoint protection.

The TI-LFA local SRLG-disjoint protection lab is concluded, and the next section summarizes the chapter.

Summary

SRLG refers to a set of links on a router that share a common risk at the transmission layer, such as through shared fiber or transmission equipment.

This means that if one link fails due to an issue in that transmission medium, all links sharing it will also fail. If both the SR-MPLS primary path and the TI-LFA backup path utilize the same transmission medium, fast reroute mechanisms may not function properly.

The TI-LFA local SRLG-disjoint protection method resolves this issue by ensuring that the backup path avoids SRLG links.

In this lab, the P2-P3 and P2-P7 links were configured as part of the same SRLG named SRLG-100 and assigned an SRLG value of 100. This configuration informed the router P2 that these links shared the same risk.

On the router P2, SRLG protection was activated under the router's IS-IS protocol similarly to node protection in the previous chapter, by assigning a non-zero index value of 100 to it.

The RIB, FIB, and LFIB outputs, along with the backup path traceroute, confirmed that TI-LFA successfully provided SRLG protection by avoiding the P2-P7 link for routes using the primary path through P2-P3.

To provide local SRLG-disjoint protection, TI-LFA excluded the P2-P7 link for all destinations reachable via the P2-P3 link during backup path calculation. A repair tunnel was established along the computed post-convergence path to serve as the backup path, ensuring TI-LFA local SRLG-disjoint protection.

The next chapter explores TI-LFA Global Weighted SRLG protection.

References

The following are the key references cited throughout this chapter. These documents provide additional insights and technical details on MPLS, Fast Reroute, and Segment Routing, offering further reading for those interested in exploring these topics in depth.

- [RFC7490] Bryant, S., Filsfils, C., Previdi, S., Shand, M., and N. So, "Remote Loop-Free Alternate (LFA) Fast Reroute (FRR)", RFC 7490, DOI 10.17487/RFC7490, April 2015, <https://www.rfc-editor.org/info/rfc7490>.

- [RFC7916] Litkowski, S., Ed., Decraene, B., Filsfils, C., Raza, K., Horneffer, M., and P. Sarkar, "Operational Management of Loop-Free Alternates", RFC 7916, DOI 10.17487/RFC7916, July 2016, <https://www.rfc-editor.org/info/rfc7916>.

- [RFC8029] Kompella, K., Swallow, G., Pignataro, C., Ed., Kumar, N., Aldrin, S., and M. Chen, "Detecting Multiprotocol Label Switched (MPLS) Data-Plane Failures", RFC 8029, DOI 10.17487/RFC8029, March 2017, <https://www.rfc-editor.org/info/rfc8029>.

- [I-D.ietf-rtgwg-segment-routing-ti-lfa] Bashandy, A., Litkowski, S., Filsfils, C., Francois, P., Decraene, B., and D. Voyer, "Topology Independent Fast Reroute using Segment Routing", Work in Progress, Internet-Draft, draft-ietf-rtgwg-segment-routing-ti-lfa-13, 16 January 2024, <https://datatracker.ietf.org/doc/html/draft-ietf-rtgwg-segment-routing-ti-lfa-13>.

- *[RFC5715] Shand, M. and S. Bryant, "A Framework for Loop-Free Convergence", RFC 5715, DOI 10.17487/RFC5715, January 2010, <https://www.rfc-editor.org/info/rfc5715>.*

- *[RFC6571] Filsfils, C., Ed., Francois, P., Ed., Shand, M., Decraene, B., Uttaro, J., Leymann, N., and M. Horneffer, "Loop-Free Alternate (LFA) Applicability in Service Provider (SP) Networks", RFC 6571, DOI 10.17487/RFC6571, June 2012, <https://www.rfc-editor.org/info/rfc6571>.*

12

Lab 11 – TI-LFA Global Weighted SRLG Protection

The previous chapter explored TI-LFA SRLG protection in a local SRLG-disjoint scenario, where a router is aware only of the SRLGs of its local links. In situations where another link, not directly connected to the router, is part of the same SRLG and potentially along the backup path, the TI-LFA Local SRLG-disjoint protection method fails to exclude that link during the calculation of the post-convergence backup path. This can result in a backup path that fails simultaneously with the primary path, as the router unknowingly includes an SRLG from a remote route. This occurs because routers in the network lack information about each other's SRLGs.

In a segment routing MPLS network, the link-state routing protocol (in this case, IS-IS), carries link-state information and propagates SRLG details as another **TLV** (**Type-Length-Value**), ensuring that all routers are aware of both local and remote SRLGs.

TI-LFA Global Weighted SRLG protection utilizes this information, providing a backup path that excludes both local and remote SRLG links when calculating the post-convergence path.

As depicted in the topology diagram, in this scenario, the P7-P3 link shares the same SRLG as P2-P3 and P2-P7.

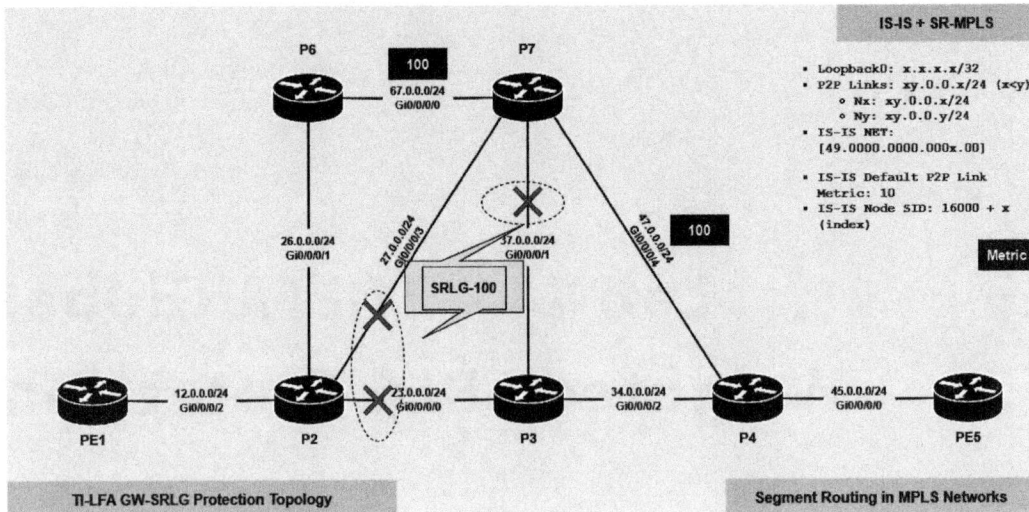

Figure 12.1 – TI-LFA Global Weighted SRLG protection

This means that if the P2-P3 link fails, the other two would also go down. In the previous lab, the TI-LFA Local SRLG-disjoint protection path from P2 to PE5 (P2-P6-P7-P3-P4-PE5) traversed the P7-P3 link, meaning that in the current lab, the backup path would also fail, leading to traffic blackholing until IGP convergence completes and a new primary path to PE5 is established on P2.

Objective

This lab continues from the previous chapter. The objective is to explore a scenario where TI-LFA Global Weighted SRLG protection provides a backup path that avoids both local and remote SRLGs in a network, enabling fast rerouting of traffic from the PLR to the destination. This is achieved through the following outlined tasks:

1. Apply the configuration to make the P7-P3 link part of the same SRLG as in the previous lab

2. Advertise the SRLG information to all the routers in the network through the IS-IS LSPDU

3. Request Global Weighted SRLG protection on P2, against the P2-P3 link failure

4. Verify that the backup path is computed and added to the RIB, FIB, and LFIB of P2, as requested in this lab.

Let's begin with the required configuration.

Configuration

The following configuration on the P2 and P7 routers is required for the Global Weighted SRLG protection scenario explored in this lab:

- The following configuration enables Global Weighted SRLG protection, which considers both local and remote SRLGs when calculating the TI-LFA backup path:

P2

```
router isis IGP
address-family ipv4 unicast
  fast-reroute per-prefix srlg-protection weighted-global
```

The Local SRLG-disjoint protection is already configured on the P2 router.

- The following configuration adds the P7–P3 link to the same SRLG group, SRLG-100, as configured on the P2 router. Furthermore, it advertises the SRLG information as a link attribute in the IS-IS LSPDU. This information is propagated as an **Application Specific Link Attribute** (**ASLA**) encoded in TLV 238, as defined in RFC 9479, *IS-IS Application-Specific Link Attributes*:

P7

```
srlg
interface GigabitEthernet0/0/0/1
  name SRLG-100
!
name SRLG-100
  value 100

router isis IGP
address-family ipv4 unicast
  advertise application lfa link-attributes srlg
```

For the ease of this lab, and since only the forward path from P2 to PE5 is being examined, the SRLG information is not added on the far side of the link and only locally on the routers. In production networks, both routers at the ends of a link need to have the same SRLG configuration.

The preceding configuration will result in a change to the backup path, as observed in the previous lab, and this change will be verified in the next section.

Verification

There were two changes made to the topology – the addition of SRLG to the P7 router and the activation of Global Weighted SRLG protection on the P2 router, which are verified next.

Verify P7 advertises SRLG as TLV 238

The following output confirms that TLV 238 contains the SRLG information in the IS-IS LSPDU. It is important to note that this differs from TLV 138 SRLG, which is utilized by RSVP-TE and the legacy mode of ASLA but is outside the scope of this discussion:

```
RP/0/RP0/CPU0:P2#show isis database verbose internal P7 | utility egrep -A9
code:238
   TLV code:238 length:26
     Application Specific SRLG: P3.00
       L flag: 0, SA-Length 1, UDA-Length 1
       Standard Applications: 0x20 LFA
       User Defined Applications: 0x20 LFA
       Sub-TLV Length: 10
         SubTLV code:4 length:8
           Local Interface ID: 8, Remote Interface ID: 8
       SRLGs:
         [0]: 100
RP/0/RP0/CPU0:P2#
```

The relevant aspects of the outputs are explained as follows:

- TLV code 238 is designated to identify the **Application-Specific SRLG TLV** used to advertise SRLG information within the IS-IS LSPDU.

- Application Specific SRLG: P3.00 indicates that the P7 router associates the SRLG with the P3 router on the far side of the advertised link.

- The Standard Applications: 0x20 LFA and User Defined Applications: 0x20 LFA attributes signify that those applications, such as LFAs, including TI-LFA, can utilize this SRLG information to compute backup paths.

- SubTLV code 4 includes details such as local and remote interface IDs specific to the P7-P3 SRLG link, facilitating precise identification.

- The SRLG 100 value serves as a common identifier, grouping multiple links together under the same SRLG, indicating shared risk characteristics.

The P7 router propagates this information via IS-IS throughout the network, enabling other routers to understand and utilize the SRLG details for backup path calculation.

The preceding output confirms that the P2 router has learned the SRLG value associated with the P7-P3 link from the IS-IS LSPDU, transmitted by the P7 router. The next step involves validating the TI-LFA backup path using this updated information.

Backup path in the RIB

The following output differs slightly compared to the output from the previous lab:

```
RP/0/RP0/CPU0:P2#show isis fast-reroute detail 5.5.5.5/32


L2 5.5.5.5/32 [30/115] Label: 16005, medium priority
    Installed Jan 01 07:48:18.259 for 00:00:26
      via 23.0.0.3, GigabitEthernet0/0/0/0, Label: 16005, P3, SRGB Base: 16000,
Weight: 0
        Backup path: TI-LFA (srlg), via 26.0.0.6, GigabitEthernet0/0/0/1 P6,
SRGB Base: 16000, Weight: 0, Metric: 220
        Backup tunnel: tunnel-te32769
          P node: P6.00 [6.6.6.6], Label: ImpNull
          Q node: P7.00 [7.7.7.7], Label: 24023
          Q node: P4.00 [4.4.4.4], Label: 24023
          Prefix label: 16005
          Backup-src: PE5.00
        P: No, TM: 220, LC: No, NP: No, D: No, SRLG: Yes
      src PE5.00-00, 5.5.5.5, prefix-SID index 5, R:0 N:1 P:0 E:0 V:0 L:0,
Alg:0
RP/0/RP0/CPU0:P2#
```

In this instance, the backup path is a backup tunnel, automatically signaled when the labels required for the requested backup path exceed the maximum supported label count of 3. In this case, the number of labels in the label stack of the backup path is 4.

The MPLS Traffic Engineering lab was essential to ensure that, if needed, the router could signal an auto tunnel.

If no auto-tunnel range is configured, the router allocates a dynamic tunnel ID, in this case, 32769, which is applicable in this case.

The TI-LFA path computation method has resulted in a backup path that excludes both local and remote SRLG links. This exclusion has caused the label stack to exceed three labels, triggering the signaling of the auto-tunnel. The auto-tunnel then utilizes the path computed by TI-LFA for its operation:

```
RP/0/RP0/CPU0:P2#show route 5.5.5.5/32

Routing entry for 5.5.5.5/32
  Known via "isis IGP", distance 115, metric 30, labeled SR, type level-2
  Installed Jan  1 07:48:18.260 for 00:00:45
  Routing Descriptor Blocks
    directly connected, via tunnel-te32769, Backup (Local-LFA)
      Route metric is 220
    23.0.0.3, from 5.5.5.5, via GigabitEthernet0/0/0/0, Protected
      Route metric is 30
```

```
     No advertising protos.
RP/0/RP0/CPU0:P2#
```

> **Note**
>
> The preceding output indicates TI-LFA (srlg), which essentially means TI-LFA
> (srlg+link), since link protection is enabled by default and cannot be disabled.

The routing table classifies this as a Local-LFA entry because the tunnel interface is directly connected to the P2 router.

Backup path in the FIB

The same information is downloaded in the FIB from the RIB:

```
RP/0/RP0/CPU0:P2#show cef 5.5.5.5/32 brief
5.5.5.5/32, version 1007, labeled SR, internal 0x1000001 0x8310 (ptr
0xe7ac838) [1], 0x600 (0xda5df48), 0xa28 (0x2218eb50)
Updated Jan  1 07:48:18.265
remote adjacency to GigabitEthernet0/0/0/0
Prefix Len 32, traffic index 0, precedence n/a, priority 1
   via 0.0.0.0/32, tunnel-te32769, 9 dependencies, weight 0, class 0, backup
(Local-LFA) [flags 0x300]
    path-idx 0 NHID 0x0 [0x22033410 0x0]
    next hop 0.0.0.0/32
    local adjacency
     local label 16005      labels imposed {16005}
   via 23.0.0.3/32, GigabitEthernet0/0/0/0, 8 dependencies, weight 0, class 0,
protected [flags 0x400]
    path-idx 1 bkup-idx 0 NHID 0x0 [0xd674540 0x0]
    next hop 23.0.0.3/32
     local label 16005      labels imposed {16005}
RP/0/RP0/CPU0:P2#
```

The FIB similarly indicates no additional label on the tunnel interface.

Backup path in the LFIB

The LFIB also confirms the absence of an additional label on the tunnel interface:

```
RP/0/RP0/CPU0:P2#show mpls forwarding prefix 5.5.5.5/32 detail
Sun Dec 31 14:37:09.606 UTC
Local  Outgoing    Prefix               Outgoing       Next Hop        Bytes
Label  Label       or ID                Interface                      Switched
------ ----------- -------------------- -------------- --------------- ----------
 --
```

```
16005   16005         SR Pfx (idx 5)      Gi0/0/0/0     23.0.0.3        0
     Updated: Dec 31 14:30:57.454
     Path Flags: 0x400 [ BKUP-IDX:0 (0xd676af0) ]
     Version: 3782, Priority: 1
     Label Stack (Top -> Bottom): { 16005 }
     NHID: 0x0, Encap-ID: N/A, Path idx: 1, Backup path idx: 0, Weight: 0
     MAC/Encaps: 4/8, MTU: 1500
     Outgoing Interface: GigabitEthernet0/0/0/0 (ifhandle 0x01000018)
     Packets Switched: 0


      16005        SR Pfx (idx
5)    tt32769       point2point     0              (!)
     Updated: Dec 31 14:30:57.454
     Path Flags: 0x300 [ IDX:0 BKUP, NoFwd ]
     Version: 3782, Priority: 1
     Label Stack (Top -> Bottom): { Unlabelled 16005 }
     NHID: 0x0, Encap-ID: N/A, Path idx: 0, Backup path idx: 0, Weight: 0
     MAC/Encaps: 0/4, MTU: 0
     Outgoing Interface: tunnel-te32769 (ifhandle 0x00000054)
     Packets Switched: 0
     (!): FRR pure backup


  Traffic-Matrix Packets/Bytes Switched: 0/0
RP/0/RP0/CPU0:P2#
```

It also shows that the backup path is recursively calculated. Initially, TI-LFA computes the post-convergence backup path excluding the specified SRLGs, resulting in an auto-tunnel interface. Subsequently, this new auto-tunnel local interface becomes the classic LFA outgoing interface.

We'll explore the tunnel interface further next.

Examining the MPLS Traffic-Eng auto-tunnel

The detailed output of the following tunnel displays the Segment-Routing Path Info details at the bottom, aligning with the list of labels observed in the preceding TI-LFA path computation output.

This is different from RSVP-TE tunnels, which use signaling. The SR-MPLS Traffic Engineering auto-tunnel does not maintain the state of the routers; instead, it encodes the entire path in the tunnel as a list of segments:

```
RP/0/RP0/CPU0:P2#show mpls traffic-eng tunnels 32769

Name: tunnel-te32769  Destination: 0.0.0.0  Ifhandle:0x34 (auto-tunnel for
ISIS IGP)
  Signalled-Name: auto_P2_t32769
  Status:
```

```
      Admin:    up Oper:   up   Path: valid   Signalling: connected

    path option (_te32769), preference 10, (verbatim Segment-Routing) type
  explicit (_te32769) (Basis for Setup)
    G-PID: 0x0800 (derived from egress interface properties)
    Bandwidth Requested: 0 kbps  CT0
    Creation Time: Mon Jan  1 07:48:17 2024 (00:02:18 ago)
  Config Parameters:
    Bandwidth:         0 kbps (CT0) Priority:  7  7 Affinity: 0x0/0xffff
    Metric Type: TE (global)
    Path Selection:
      Tiebreaker: Min-fill (default)
      Protection: any (default)
    Hop-limit: disabled
    Cost-limit: disabled
    Delay-limit: disabled
    Delay-measurement: disabled
    Path-invalidation timeout: 10000 msec (default), Action: Tear (default)
    AutoRoute: disabled  LockDown: disabled   Policy class: not set
    Forward class: 0 (not enabled)
    Forwarding-Adjacency: disabled
    Autoroute Destinations: 0
    Loadshare:         0 equal loadshares
    Auto-bw: disabled
    Auto-Capacity: Disabled:
    Path Protection: Not Enabled
    BFD Fast Detection: Disabled
    Reoptimization after affinity failure: Enabled
    SRLG discovery: Disabled
  History:
    Tunnel has been up for: 00:02:18 (since Mon Jan 01 07:48:17 UTC 2024)
    Current LSP:
      Uptime: 00:02:18 (since Mon Jan 01 07:48:17 UTC 2024)

  Segment-Routing Path Info (IGP information is not used)
    Segment0[First Hop]: 26.0.0.6, Label: -
    Segment1[ - ]: Label: 24023
    Segment2[ - ]: Label: 24023

Displayed 1 (of 2) heads, 0 (of 0) midpoints, 0 (of 0) tails
Displayed 1 up, 0 down, 0 recovering, 0 recovered heads
RP/0/RP0/CPU0:P2#
```

The relevant aspects of the preceding output for this lab are explained as follows:

- The tunnel interface name is displayed as Name: tunnel-te32769.

- This tunnel follows a labeled path rather than an IP path, indicated by `Destination: 0.0.0.0.`

- (`auto-tunnel for ISIS IGP`) confirms that it is an SR-MPLS auto-tunnel, using IS-IS as the IGP.

- `Segment-Routing Path Info` shows that the TI-LFA computed path includes constraints and differs from the standard IGP path.

- The first hop from the P2 router on this path is the P6 router, with a directly connected neighbor having an implicit null label, shown by `Segment0[First Hop]: 26.0.0.6, Label: -.`

- Subsequent segments are encoded as labels, shown by `Segment1[ - ]: Label: 24023` and `Segment2[ - ]: Label: 24023.`

This output details the auto-tunnel, highlighting the use of the MPLS traffic-engineering component in Cisco IOS XR for RSVP-TE tunnels. It has been repurposed for scenarios where the backup path label stack exceeds three labels, providing a backup path auto-tunnel through this method.

The next section focuses on verifying the backup path.

Verifying the backup path traceroute

The essential components necessary for the `nil-fec` traceroute, as derived from the RIB, FIB, or LFIB, are as follows:

- **Label stack (top to bottom)**: `3`, `24023`, `24023`, `16005`, directing traffic to P6, then P6-P7, then P7-P4, and finally, PE5

- **Outgoing interface**: `GigabitEthernet0/0/0/1`, toward P6

- **Next-hop IP address of the outgoing interface**: `26.0.0.6`, to P6

The following output traces the backup path through all the hops, as per the requested label stack, outgoing interface, and the next-hop, which was computed by the TI-LFA in the previous output:

```
RP/0/RP0/CPU0:P2#traceroute sr-mpls nil-fec labels 3,24023,24023,16005 output
interface gigabitEthernet 0/0/0/1 nexthop 26.0.0.6

Tracing MPLS Label Switched Path with Nil FEC with labels
[3,24023,24023,16005], timeout is 2 seconds

Codes: '!' - success, 'Q' - request not sent, '.' - timeout,
  'L' - labeled output interface, 'B' - unlabeled output interface,
  'D' - DS Map mismatch, 'F' - no FEC mapping, 'f' - FEC mismatch,
  'M' - malformed request, 'm' - unsupported tlvs, 'N' - no rx label,
  'P' - no rx intf label prot, 'p' - premature termination of LSP,
  'R' - transit router, 'I' - unknown upstream index,
  'X' - unknown return code, 'x' - return code 0
```

```
Type escape sequence to abort.

  0 26.0.0.2 MRU 1500 [Labels: implicit-null/24023/24023/16005/explicit-null
Exp: 0/0/0/0/0]
L 1 26.0.0.6 MRU 1500 [Labels: implicit-null/24023/16005/explicit-null Exp:
0/0/0/0] 13 ms
L 2 67.0.0.7 MRU 1500 [Labels: implicit-null/16005/explicit-null Exp: 0/0/0] 9
ms
L 3 47.0.0.4 MRU 1500 [Labels: implicit-null/explicit-null Exp: 0/0] 10 ms
! 4 45.0.0.5 11 ms
RP/0/RP0/CPU0:P2#
```

The successful traceroute confirms that in the event of a P2–P3 link failure on the P2 router, the TI-LFA path uses the aforementioned label stack, outgoing interface, and next hop to successfully fast-reroute traffic through the P6 router, then via the P6–P7 link, followed by the P7–P4 link, and finally, to the destination. This rerouting successfully avoids both local and remote SRLGs, thereby providing Global Weighted SRLG protection.

The next section elaborates on how TI-LFA incorporated the additional requirement into the backup path calculation.

Understanding backup path calculation

The following topology diagram describes the backup path calculation.

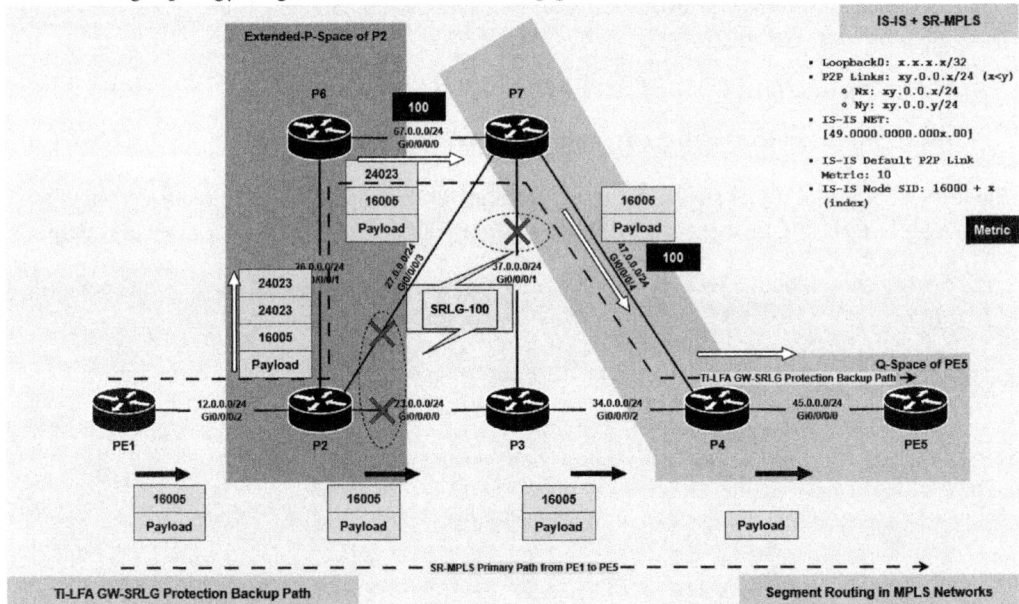

Figure 12.2 – TI-LFA Global Weighted SRLG protection primary and backup path

The following explanation details the computation and selection of the primary path and backup path.

Primary path

The primary path from P2 to PE5 goes via P3, with the end-to-end path, `cost of 30` (P2 - P3 - P4 - PE5), making it the shortest path.

Backup path

The backup path is classified as `TI-LFA (srlg)` because SRLG protection is explicitly requested in the P2 router configuration

TI-LFA

The post-convergence path in the scenario, including the local `srlg-disjoint` request, would be P2 - P6 - P7 - P4 - PE5, with a cost of 220.

In addressing the link failure on the P2–P3 link, TI-LFA provides SRLG protection by collecting the following information:

- **P-Space of P2 along the post-convergence path**: The group of routers on the post-convergence path, reachable from P2 through the shortest path tree, excluding the traversal of the P2–P3 link, the P2–P7 link, and the P7–P3 link, forms the P-Space of P2.

  In this topology, that router is P6.

- **Extended P-Space of P2 along the post-convergence path**: The extended P-Space of P2 is the combination of P2's P-Space and the P-Space of P2's neighbors along the post-convergence path. This can be determined by calculating the shortest path tree, rooted at each of P2's neighbors, and removing the subtree accessed via the P2–P3 link, the P2–P7 link, and the P7–P3 link (including the ECMP nodes).

  There are no routers in the P-Space of P6.

  Therefore, the extended P-Space of P2 along the post-convergence path includes only P6.

- **Q-Space of PE5 along the post-convergence path**: The group of routers along the post-convergence path, through which the PE5 router can be reached via the shortest path tree without traversing the P2–P3 link, the P2–P7 link, and the P7 – P3 link, forms the Q-Space of PE5.

  The Q-Space of PE5 comprises P7 and P4.

The TI-LFA repair path is a sequence of segments (a stack of labels) along the post-convergence path on the outgoing interface. This path is determined by the intersection of the extended P-Space of P2 and the Q-Space of PE5.

There is no overlap between the extended P-Space and Q-Space in this scenario. The further downstream P node is P6, the adjacent Q node to it is P7, and after incorporating the SRLG constraint, an additional Q node, P4, adjacent to P7, is necessary.

The endpoint for the repair tunnel needs to be P4 along the post-convergence path. To achieve this, P2 encodes additional information in the label stack, instructing P6 to use the adjacency toward P7, and P7 to use the adjacency toward P4, ensuring a loop-free alternate.

The preceding explanation is also evident in the verification output.

Thus, the TI-LFA backup path ensures loop-free forwarding with Global Weighted SRLG protection.

The next section restores the lab topology for the next chapter.

> **Note**
>
> Due to the topology limitations, the next chapter does not utilize Global Weighted SRLG but only a local SRLG-disjoint, which remains configured from the previous chapter.

Restoring the topology

Apply the following configuration to remove only the Global Weighted SRLG protection from the P2 router:

P2

```
router isis IGP
address-family ipv4 unicast
  no fast-reroute per-prefix srlg-protection weighted-global
```

As Global Weighted SRLG protection is removed from P2, it's not necessary to stop advertising `Application-Specific SRLG TLV` carrying SRLG information on P7 to the entire network, but it is done for completeness, as shown here.

The following configuration would stop the advertisement of the local SRLG, configured on the P7 router, in the IS-IS router through LSPDUs:

P7

```
no router isis IGP address-family ipv4 unicast advertise application lfa link-
attributes srlg
```

Now that we've explored both SRLG scenarios, let's summarize what we have learned in this chapter.

Summary

The chapter explored **SRLG (Shared Risk Link Group)** scenarios, commonly found in large-scale production service provider networks, spanning across regions and countries. These networks often encounter SRLG links that extend beyond direct connections to routers, reaching into remote areas.

The TI-LFA Local SRLG-disjoint backup path is designed to exclude local SRLG links when recalculating paths post-convergence, aiming to fast-reroute traffic during failures. However, if a remote link along this backup path shares the same SRLG, the backup path may fail concurrently with the primary path, despite being installed in the RIB, FIB, and LFIB.

This issue arises due to the lack of SRLG information exchange among remote routers in a network, as highlighted in the problem statement involving the P7–P3, P2–P3, and P2–P7 links from the previous chapter.

The solution involved two main steps – first, advertising SRLG information from the P7 router P7 throughout a network, using the IS-IS link state routing protocol and TLV 238 for application-specific SRLG TLV. This allowed all routers, including P2, to learn and store this information in their LSDB.

Secondly, on the P2 router, implementing TI-LFA Global Weighted SRLG protection enabled consideration of both local and remote SRLGs when calculating backup paths. By excluding links that share the risk value 100 (such as P2–P3, P2–P7, and P7–P3) from the computation, P2 can offer a backup path through the remaining links post-convergence.

Interestingly, the backup path offered exceeded the maximum label limit of three, prompting MPLS traffic engineering on the P2 router to activate an SR-MPLS auto-tunnel backup. Unlike RSVP-TE auto-tunnels, this SR-MPLS auto-tunnel carries its own labeled path information without signaling to other routers.

The verification steps included examining the RIB, FIB, and LFIB output, along with traceroute analysis to confirm the backup path's derivation. Finally, the topology was restored to prepare for the requirements of the next chapter.

The next chapter explores the TI-LFA Node+SRLG protection scenario.

References

Below are the key references cited throughout this chapter. These documents provide additional insights and technical details on MPLS, Fast Reroute, and Segment Routing, offering further reading for those interested in exploring these topics in depth.

- *[RFC7490] Bryant, S., Filsfils, C., Previdi, S., Shand, M., and N. So, "Remote Loop-Free Alternate (LFA) Fast Reroute (FRR)", RFC 7490, DOI 10.17487/RFC7490, April 2015, <https://www.rfc-editor.org/info/rfc7490>.*

- *[RFC7916] Litkowski, S., Ed., Decraene, B., Filsfils, C., Raza, K., Horneffer, M., and P. Sarkar, "Operational Management of Loop-Free Alternates", RFC 7916, DOI 10.17487/RFC7916, July 2016, <https://www.rfc-editor.org/info/rfc7916>.*

- *[RFC8029] Kompella, K., Swallow, G., Pignataro, C., Ed., Kumar, N., Aldrin, S., and M. Chen, "Detecting Multiprotocol Label Switched (MPLS) Data-Plane Failures", RFC 8029, DOI 10.17487/RFC8029, March 2017, <https://www.rfc-editor.org/info/rfc8029>.*

- *[I-D.ietf-rtgwg-segment-routing-ti-lfa] Bashandy, A., Litkowski, S., Filsfils, C., Francois, P., Decraene, B., and D. Voyer, "Topology Independent Fast Reroute using Segment Routing", Work in Progress, Internet-Draft, draft-ietf-rtgwg-segment-routing-ti-lfa-13, 16 January 2024, <https://datatracker.ietf.org/doc/html/draft-ietf-rtgwg-segment-routing-ti-lfa-13>.*

- *[RFC5715] Shand, M. and S. Bryant, "A Framework for Loop-Free Convergence", RFC 5715, DOI 10.17487/RFC5715, January 2010, <https://www.rfc-editor.org/info/rfc5715>.*

- *[RFC6571] Filsfils, C., Ed., Francois, P., Ed., Shand, M., Decraene, B., Uttaro, J., Leymann, N., and M. Horneffer, "Loop-Free Alternate (LFA) Applicability in Service Provider (SP) Networks", RFC 6571, DOI 10.17487/RFC6571, June 2012, <https://www.rfc-editor.org/info/rfc6571>.*

- *[RFC9479] Ginsberg, L., Psenak, P., Previdi, S., Henderickx, W., and J. Drake, "IS-IS Application-Specific Link Attributes", RFC 9479, DOI 10.17487/RFC9479, October 2023, <https://www.rfc-editor.org/info/rfc9479>.*

13
Lab 12 – TI-LFA Node + SRLG Protection

In this book, the exploration of TI-LFA has delved deeply into node protection and **Shared Risk Link Group (SRLG)** protection as standalone strategies. However, in production network environments, the demand often arises for comprehensive fast reroute solutions that can simultaneously provide resilience against both node failures and shared risk link failures.

TI-LFA is a mechanism designed to address these needs by offering a combined approach where topology permits. Node protection in TI-LFA ensures that in the event of a node failure along the primary path, traffic is rerouted using a backup path that avoids the failed node, thereby minimizing disruption. On the other hand, SRLG protection extends this capability by considering not just individual link failures but also the risk of failures associated with groups of links that share common risk factors, such as transmission resource dependencies.

In real-world scenarios, TI-LFA's integrated Node + SRLG protection proves invaluable. For instance, in large-scale service provider networks where maintaining high availability is critical, TI-LFA can dynamically reroute traffic around both node failures and SRLG failures; the **Point of Local Repair (PLR)** router calculates alternative paths that not only circumvent the node around the failed link but also consider SRLGs associated with the affected link. This ensures continuous service delivery without relying solely on IGP convergence.

Objective

Continuing from the previous chapter, the topology already features local SRLG-disjoint protection. This lab's objective is to explore a scenario where TI-LFA is configured to provide a node and SRLG-protected fast reroute backup path. The tasks outlined next accomplish this goal:

1. Apply the `ti-lfa` node protection configuration on P2
2. Verify that the `node+srlg` backup path is computed and added to the RIB, FIB, and LFIB of P2.

The topology described in *Figure 13.1* illustrates the TI-LFA Node + SRLG protection configuration.

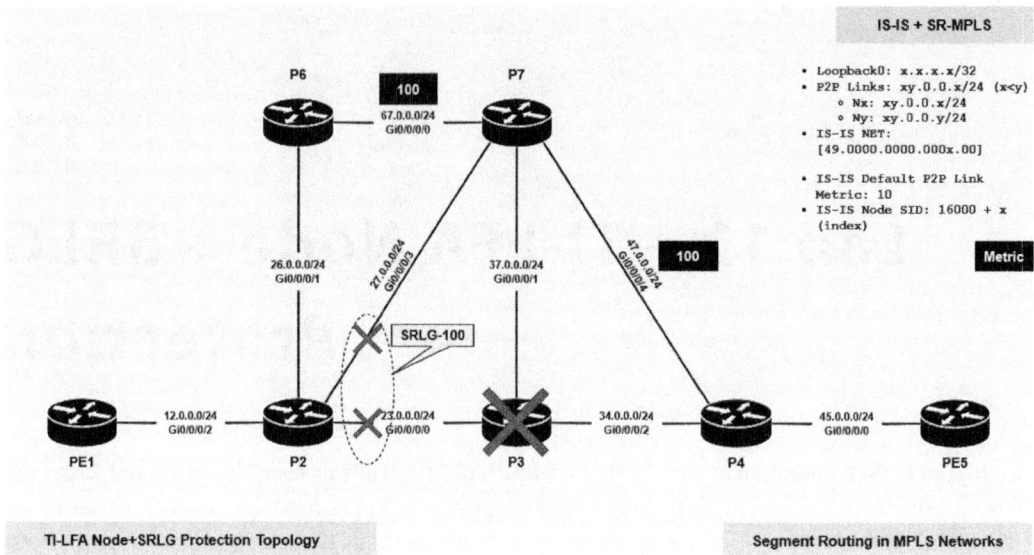

Figure 13.1 – TI-LFA Node + SRLG protection

At the end of the previous chapter, the configuration on router P2 was changed to remove only the Global Weighted SRLG protection, while the local SRLG-disjoint protection remains configured.

Configuration

Activate TI-LFA node protection on P2 as depicted next.

By activating node protection alongside the existing local SRLG-disjoint protection, router P2 is configured to provide Node + SRLG protection wherever available:

P2

```
router isis IGP
address-family ipv4 unicast
  fast-reroute per-prefix tiebreaker node-protecting index 200
```

The same index value of 200 used earlier in the node protection in *Chapter 10* is also utilized here.

The next section verifies the availability of the node+SRLG protection.

Verification

The backup path shown next in the RIB, FIB, and LFIB confirms that the PLR router P2 provides node as well as SRLG protection.

Backup path in the RIB

The following output confirms that the provided fast reroute path includes node+SRLG protection in the event of a failure on the P2–P3 link:

```
RP/0/RP0/CPU0:P2#show isis fast-reroute detail 5.5.5.5/32


L2 5.5.5.5/32 [30/115] Label: 16005, medium priority
    Installed Jan 01 07:54:10.977 for 00:00:17
      via 23.0.0.3, GigabitEthernet0/0/0/0, Label: 16005, P3, SRGB Base: 16000,
Weight: 0
        Backup path: TI-LFA (node+srlg), via 26.0.0.6, GigabitEthernet0/0/0/1
P6, SRGB Base: 16000, Weight: 0, Metric: 220
      Backup tunnel: tunnel-te32769
        P node: P6.00 [6.6.6.6], Label: ImpNull
        Q node: P7.00 [7.7.7.7], Label: 24023
        Q node: P4.00 [4.4.4.4], Label: 24023
        Prefix label: 16005
        Backup-src: PE5.00
      P: No, TM: 220, LC: No, NP: Yes, D: No, SRLG: Yes
    src PE5.00-00, 5.5.5.5, prefix-SID index 5, R:0 N:1 P:0 E:0 V:0 L:0,
Alg:0
RP/0/RP0/CPU0:P2#
```

The number of labels exceeded the maximum supported limit of three. Therefore, as described earlier in *Chapter 10*, a **Segment Routing Auto Tunnel (SRAT)** is initialized.

Note

The preceding output indicates TI-LFA (node+srlg), which essentially means TI-LFA (node+srlg+link) since link protection is enabled by default and cannot be disabled.

Backup path in the FIB

As noted earlier in the book, when an SRAT is initialized, the FIB utilizes the tunnel interfaces as outgoing backup interfaces:

```
RP/0/RP0/CPU0:P2#show cef 5.5.5.5/32 brief
5.5.5.5/32, version 1035, labeled SR, internal 0x1000001 0x8310 (ptr
0xe7ac838) [1], 0x600 (0xda5df48), 0xa28 (0x2218f5f8)
Updated Jan  1 07:54:10.983
remote adjacency to GigabitEthernet0/0/0/0
Prefix Len 32, traffic index 0, precedence n/a, priority 1
   via 0.0.0.0/32, tunnel-te32769, 9 dependencies, weight 0, class 0, backup
(Local-LFA) [flags 0x300]
     path-idx 0 NHID 0x0 [0x22033410 0x0]
```

```
   next hop 0.0.0.0/32
   local adjacency
     local label 16005        labels imposed {16005}
   via 23.0.0.3/32, GigabitEthernet0/0/0/0, 8 dependencies, weight 0, class 0,
protected [flags 0x400]
     path-idx 1 bkup-idx 0 NHID 0x0 [0xd673cd0 0x0]
   next hop 23.0.0.3/32
     local label 16005        labels imposed {16005}
RP/0/RP0/CPU0:P2#
```

In such scenarios, the backup mechanism also reflects Local-LFA, since the tunnel interface is directly connected to the PLR.

Backup path in the LFIB

The information is downloaded into the LFIB:

```
RP/0/RP0/CPU0:P2#show mpls forwarding prefix 5.5.5.5/32
Local   Outgoing    Prefix              Outgoing        Next Hop          Bytes
Label   Label       or ID               Interface                         Switched
------  ----------  ------------------- ------------    ---------------   ----------
--
16005   16005       SR Pfx (idx 5)      Gi0/0/0/0       23.0.0.3          0
        16005       SR Pfx (idx
5)      tt32769         point2point     0                (!)
RP/0/RP0/CPU0:P2#
```

The outgoing interface for the backup path is now the tunnel interface, which will be examined in the next section.

Examining the MPLS traffic-eng auto tunnel

The operational state of the initialized SRAT is shown next. Segment-Routing Path Info depicts the segments this tunnel has encapsulated as labels to reach the destination:

```
RP/0/RP0/CPU0:P2#show mpls traffic-eng tunnels 32769

Name: tunnel-te32769  Destination: 0.0.0.0  Ifhandle:0x34 (auto-tunnel for
ISIS IGP)
  Signalled-Name: auto_P2_t32769
  Status:
    Admin:    up Oper:    up   Path: valid   Signalling: connected

    path option (_te32769), preference 10, (verbatim Segment-Routing) type
explicit (_te32769) (Basis for Setup)
    G-PID: 0x0800 (derived from egress interface properties)
```

```
      Bandwidth Requested: 0 kbps   CT0
      Creation Time: Mon Jan  1 07:48:17 2024 (00:07:53 ago)
   Config Parameters:
      Bandwidth:          0 kbps (CT0) Priority:  7  7 Affinity: 0x0/0xffff
      Metric Type: TE (global)
      Path Selection:
        Tiebreaker: Min-fill (default)
        Protection: any (default)
      Hop-limit: disabled
      Cost-limit: disabled
      Delay-limit: disabled
      Delay-measurement: disabled
      Path-invalidation timeout: 10000 msec (default), Action: Tear (default)
      AutoRoute: disabled  LockDown: disabled   Policy class: not set
      Forward class: 0 (not enabled)
      Forwarding-Adjacency: disabled
      Autoroute Destinations: 0
      Loadshare:          0 equal loadshares
      Auto-bw: disabled
      Auto-Capacity: Disabled:
      Path Protection: Not Enabled
      BFD Fast Detection: Disabled
      Reoptimization after affinity failure: Enabled
      SRLG discovery: Disabled
   History:
      Tunnel has been up for: 00:07:53 (since Mon Jan 01 07:48:17 UTC 2024)
      Current LSP:
        Uptime: 00:07:53 (since Mon Jan 01 07:48:17 UTC 2024)

   Segment-Routing Path Info (IGP information is not used)
      Segment0[First Hop]: 26.0.0.6, Label: -
      Segment1[ - ]: Label: 24023
      Segment2[ - ]: Label: 24023

Displayed 1 (of 1) heads, 0 (of 0) midpoints, 0 (of 0) tails
Displayed 1 up, 0 down, 0 recovering, 0 recovered heads
RP/0/RP0/CPU0:P2#
```

The first hop has no label due to implicit null. Thereafter, the segments reflect the same as shown in the RIB previously.

The next section validates this backup path.

Verifying the backup path traceroute

The essential components necessary for the `nil-fec` traceroute, as derived from the RIB, FIB, or LFIB, are as follows:

- **Label stack (top to bottom)**: 3, 24023, 24023, 16005, directing traffic to P6, then P6–P7, then P7–P4, and finally to PE5

- **Outgoing interface**: GigabitEthernet0/0/0/1, toward P6

- **Next-hop IP address of outgoing interface**: 26.0.0.6, to P6

The following output traces the backup path through all the hops, as per the requested label stack, outgoing interface, and the next hop, which was computed by the TI-LFA in the previous output:

```
RP/0/RP0/CPU0:P2#traceroute sr-mpls nil-fec labels 3,24023,24023,16005 output
interface GigabitEthernet 0/0/0/1 nexthop 26.0.0.6

Tracing MPLS Label Switched Path with Nil FEC with labels
[3,24023,24023,16005], timeout is 2 seconds

Codes: '!' - success, 'Q' - request not sent, '.' - timeout,
   'L' - labeled output interface, 'B' - unlabeled output interface,
   'D' - DS Map mismatch, 'F' - no FEC mapping, 'f' - FEC mismatch,
   'M' - malformed request, 'm' - unsupported tlvs, 'N' - no rx label,
   'P' - no rx intf label prot, 'p' - premature termination of LSP,
   'R' - transit router, 'I' - unknown upstream index,
   'X' - unknown return code, 'x' - return code 0

Type escape sequence to abort.

  0 26.0.0.2 MRU 1500 [Labels: implicit-null/24023/24023/16005/explicit-null
Exp: 0/0/0/0/0]
L 1 26.0.0.6 MRU 1500 [Labels: implicit-null/24023/16005/explicit-null Exp:
0/0/0/0] 19 ms
L 2 67.0.0.7 MRU 1500 [Labels: implicit-null/16005/explicit-null Exp: 0/0/0]
16 ms
L 3 47.0.0.4 MRU 1500 [Labels: implicit-null/explicit-null Exp: 0/0] 19 ms
! 4 45.0.0.5 16 ms
RP/0/RP0/CPU0:P2#
```

The successful traceroute confirms that in the event of a P2–P3 link failure on router P2, the TI-LFA path utilizes the aforementioned label stack, outgoing interface, and next hop to effectively reroute traffic through router P6, then via the P6–P7 link, followed by the P7–P4 link, and finally, to the destination. This rerouting successfully bypasses the next-hop router P3 and the local SRLGs on router P2, providing node and SRLG protection.

The next section elaborates on how TI-LFA incorporated this additional requirement into the backup path calculation.

Understanding the backup path calculation

Figure 13.2 illustrates the TI-LFA Node+SRLG protection.

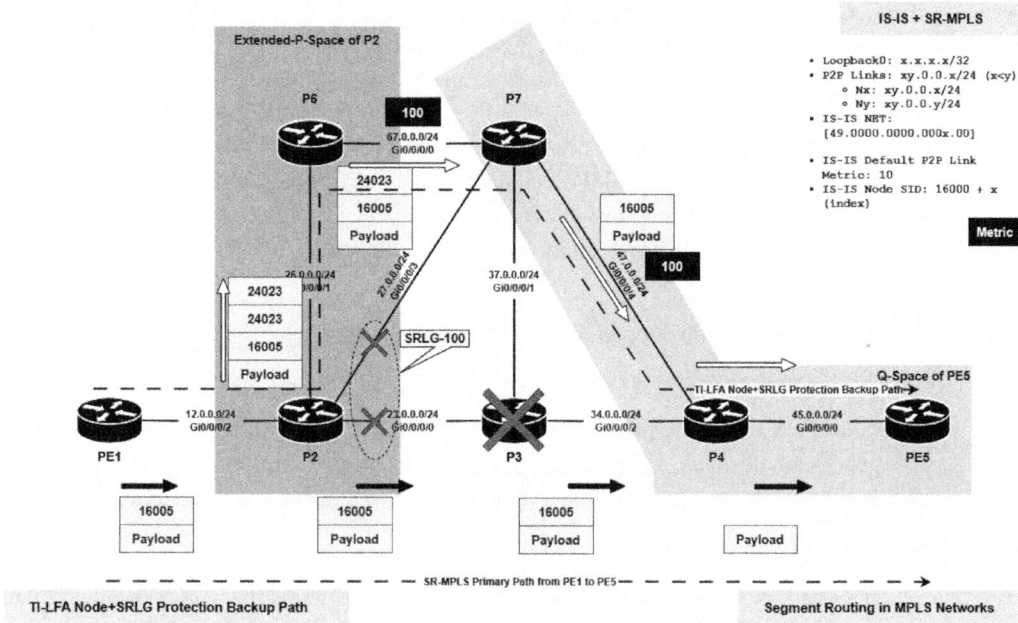

Figure 13.2 – TI-LFA Node + SRLG protection primary and backup paths

The following explanation details the computation and selection of the primary path and backup path.

Primary path

The primary path from P2 to PE5 goes via P3, with the end-to-end path cost of 30 (P2 - P3 - P4 - PE5), making it the shortest path.

Backup path

The backup path is categorized as TI-LFA (node+srlg) since both node protection and SRLG protection are explicitly requested in the P2 router configuration.

TI-LFA

The post-convergence path for this request would be P2 - P6 - P7 - P4 - PE5, with a cost of 220. In addressing the link failure on the P2–P3 link, TI-LFA provides Node+SRLG protection by collecting the following information:

- **P-Space of P2 along the post-convergence path**: The group of routers on the post-convergence path reachable from P2 through the shortest path tree, excluding the traversal of the P3 router, P2–P3 link and the P2–P7 link, forms the P-Space of P2.

  In this topology, that router is P6.

- **Extended P-Space of P2 along the post-convergence path**: The extended P-Space of P2 is the combination of P2's P-Space and the P-Space of P2's neighbors along the post-convergence path. This can be determined by calculating the shortest path tree rooted at each of P2's neighbors and removing the subtree accessed via the P3 router, P2–P3 link, and the P2–P7 link (including the ECMP nodes).

  There are no routers in the P-Space of P6.

  Therefore, the extended P-Space of P2 along the post-convergence path includes only P6.

- **Q-Space of PE5 along the post-convergence path**: The group of routers along the post-convergence path, through which router PE5 can be reached via the shortest path tree without traversing the P3 router, P2–P3 link, and the P2–P7 link, forms the Q-Space of PE5.

  The Q-Space of PE5 comprises of P7 and P4.

The TI-LFA repair path is a sequence of segments (a stack of labels) along the post-convergence path on the outgoing interface. This path is determined by the intersection of the extended P-Space of P2 and the Q-Space of PE5.

There is no overlap between the extended P-Space and Q-Space in this scenario. The further downstream P node is P6, the adjacent Q node to it is P7, and after incorporating the Node+SRLG constraint, an additional Q node, P4, adjacent to P7, is necessary.

The endpoint for the repair tunnel needs to be P4 along the post-convergence path. To achieve this, P2 encodes additional information in the label stack instructing P6 to use the adjacency toward P7, and P7 to use the adjacency toward P4, ensuring a loop-free alternate.

The preceding explanation is also evident in the verification outputs.

Thus, the TI-LFA backup path ensures loop-free forwarding with Node+SRLG protection.

With this, the chapter concludes.

Summary

This chapter combined the previously explored concepts of TI-LFA node protection and TI-LFA SRLG protection, aiming to provide a comprehensive backup path solution that integrates both protection mechanisms. When a link or node failure occurs, TI-LFA on the PLR calculates an alternative path that bypasses not only the next-hop router but also local SRLGs, ensuring continuous traffic forwarding.

Building upon the preceding chapter where Global Weighted SRLG protection was removed from router P2 while local SRLG-disjoint protection remained configured, TI-LFA node protection was subsequently enabled on this router. This configuration instructed TI-LFA on P2 to compute backup paths that incorporate both node and SRLG protection. The outputs from the RIB, FIB, and LFIB confirmed that in the event of a failure on the P2–P3 link, the rerouted path through P6, P7, and P4 successfully provided backup connectivity.

During the examination of the RIB, it was observed that the number of required labels exceeded the supported limit of three, triggering the initialization of an SRAT. This SRAT encapsulated backup path segments as labels to maintain continuity in traffic flow.

The validation of the backup path through traceroute confirmed the successful rerouting capability toward destination router PE5, leveraging information from the RIB outputs. The chapter also delved into the backup path calculation process, highlighting the familiar methodology used by TI-LFA in previous instances detailed in the book. With this exploration, the current chapter concludes, setting the stage for the subsequent chapter, which will explore TI-LFA tiebreaker scenarios.

References

Below are the key references cited throughout this chapter. These documents provide additional insights and technical details on MPLS, Fast Reroute, and Segment Routing, offering further reading for those interested in exploring these topics in depth.

- [RFC7490] Bryant, S., Filsfils, C., Previdi, S., Shand, M., and N. So, "Remote Loop-Free Alternate (LFA) Fast Reroute (FRR)", RFC 7490, DOI 10.17487/RFC7490, April 2015, <https://www.rfc-editor.org/info/rfc7490>.

- [RFC7916] Litkowski, S., Ed., Decraene, B., Filsfils, C., Raza, K., Horneffer, M., and P. Sarkar, "Operational Management of Loop-Free Alternates", RFC 7916, DOI 10.17487/RFC7916, July 2016, <https://www.rfc-editor.org/info/rfc7916>.

- [RFC8029] Kompella, K., Swallow, G., Pignataro, C., Ed., Kumar, N., Aldrin, S., and M. Chen, "Detecting Multiprotocol Label Switched (MPLS) Data-Plane Failures", RFC 8029, DOI 10.17487/RFC8029, March 2017, <https://www.rfc-editor.org/info/rfc8029>.

- *[I-D.ietf-rtgwg-segment-routing-ti-lfa] Bashandy, A., Litkowski, S., Filsfils, C., Francois, P., Decraene, B., and D. Voyer, "Topology Independent Fast Reroute using Segment Routing", Work in Progress, Internet-Draft, draft-ietf-rtgwg-segment-routing-ti-lfa-13, 16 January 2024, <https://datatracker.ietf.org/doc/html/draft-ietf-rtgwg-segment-routing-ti-lfa-13>.*

- *[RFC5715] Shand, M. and S. Bryant, "A Framework for Loop-Free Convergence", RFC 5715, DOI 10.17487/RFC5715, January 2010, <https://www.rfc-editor.org/info/rfc5715>.*

- *[RFC6571] Filsfils, C., Ed., Francois, P., Ed., Shand, M., Decraene, B., Uttaro, J., Leymann, N., and M. Horneffer, "Loop-Free Alternate (LFA) Applicability in Service Provider (SP) Networks", RFC 6571, DOI 10.17487/RFC6571, June 2012, <https://www.rfc-editor.org/info/rfc6571>.*

14

Lab 13 – TI-LFA Tiebreaker

In the previous chapter, the concept of TI-LFA node + SRLG protection was explored, demonstrating how a fast reroute backup path can be provided in networks where both node and SRLG failures are a concern. By combining these two protection mechanisms, comprehensive coverage is ensured, allowing traffic to be rerouted in various failure scenarios. This dual protection (link protection is enabled by default) is particularly valuable in large and complex production networks.

However, there are instances where TI-LFA can only offer either node protection or SRLG protection, but not both simultaneously. In such cases, the **Point of Local Repair** (**PLR**) router must prioritize one protection mechanism over the other. This is where the tiebreaker index values for each protection method, introduced in previous chapters, become crucial.

The decision to prioritize either node protection or SRLG protection is based on several factors. For example, if the underlying transmission network is subject to multiple SRLGs, prioritizing SRLG protection ensures path diversity. Conversely, if a node represents a single point of failure for multiple paths, prioritizing node protection over SRLG protection is essential to maintain network resilience.

This chapter will comprise three labs. These labs will investigate scenarios involving tiebreakers between link protection, node protection, and SRLG protection. To facilitate these scenarios, an additional router, P9, will be incorporated into the topology, as depicted in the diagram in *Figure 14.1*.

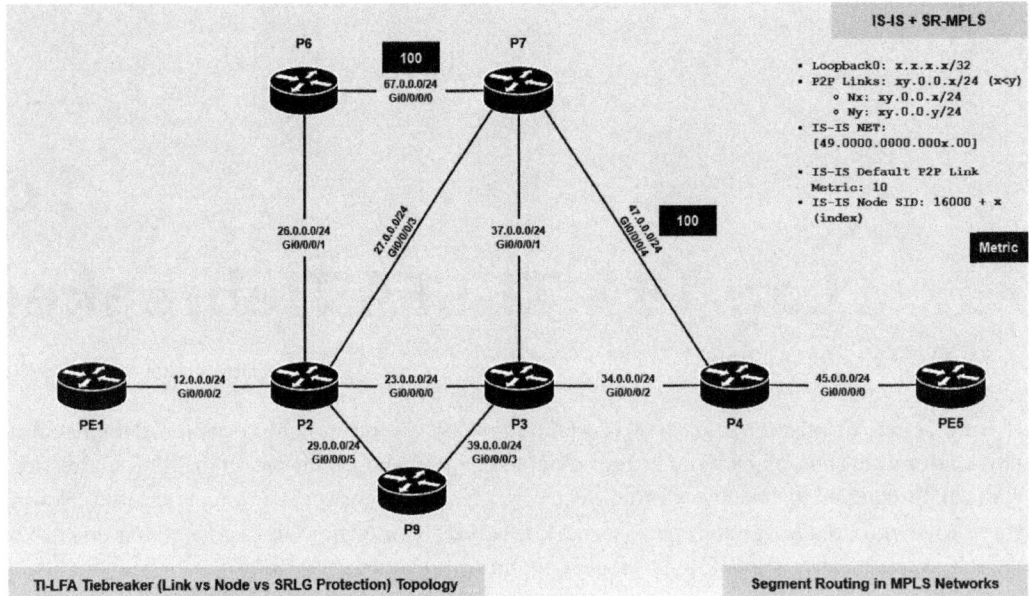

Figure 14.1 – TI-LFA tiebreaker

The configuration in the next section covers the integration of a new router, P9, into the existing SR-MPLS with TI-LFA network topology.

Preparation

Let us now implement the IP address, IS-IS, and SR-MPLS configurations in alignment with the configurations applied to the remaining routers in the topology:

P9

```
interface Loopback0
ipv4 address 9.9.9.9 255.255.255.255
!
interface GigabitEthernet0/0/0/3
ipv4 address 39.0.0.9 255.255.255.0
no shutdown
!
interface GigabitEthernet0/0/0/5
ipv4 address 29.0.0.9 255.255.255.0
no shutdown
!
```

The router P9 is connected to the router P2 and P3. The schema for each interface is defined as follows:

For `interface loopback0`, the value of x is the number associated with the router hostname:

- P9: The value of x = 9

For the following physical interface, the values of x and y are the numbers associated with the hostnames on either end of the link:

- P9-P2: The value of x = 2, y = 9 (where x < y)
- P9-P3: The value of x = 3, y = 9 (where x < y)

The IPv4 address template from *Chapter 2* is applied using the specified values.

The IS-IS configuration follows the convention defined for the NET ID and the Prefix-SID index at the beginning of the book, x = 9.

Additionally, router P9 is integrated directly into the SR-MPLS domain since LDP is no longer used:

```
router IS-IS IGP
is-type level-2-only
net 49.0000.0000.0009.00
log adjacency changes
address-family ipv4 unicast
  metric-style wide level 2
  segment-routing mpls sr-prefer
!
interface Loopback0
  passive
  address-family ipv4 unicast
   prefix-sid index 9
   !
!
```

Router P9 is connected to routers P2 and P3, with the IS-IS configuration also enabled with TI-LFA:

```
router IS-IS IGP

interface GigabitEthernet0/0/0/3
  circuit-type level-2-only
  point-to-point
  hello-padding disable
  address-family ipv4 unicast
   fast-reroute per-prefix
   fast-reroute per-prefix ti-lfa
   metric 10
   !
```

```
!
interface GigabitEthernet0/0/0/5
  circuit-type level-2-only
  point-to-point
  hello-padding disable
  address-family ipv4 unicast
   fast-reroute per-prefix
   fast-reroute per-prefix ti-lfa
   metric 10
   !
 !
 !
mpls oam
```

Finally, the MPLS data plane failure detection tools are activated to validate the forwarding plane through P9, completing the configuration.

The following configuration on P2 supports integration with P9 by configuring the IPv4 address on the link connected to P9 and activating IS-IS on that interface to include it in the IGP:

P2

```
interface GigabitEthernet0/0/0/5
ipv4 address 29.0.0.2 255.255.255.0
no shutdown
!
router IS-IS IGP
interface GigabitEthernet0/0/0/5
  circuit-type level-2-only
  point-to-point
  hello-padding disable
  address-family ipv4 unicast
   fast-reroute per-prefix
   fast-reroute per-prefix ti-lfa
   metric 10
```

The following configuration on P3 supports integration with P9 by configuring the IPv4 address on the link connected to P9 and activating IS-IS on that interface to include it in the IGP:

P3

```
interface GigabitEthernet0/0/0/3
ipv4 address 39.0.0.3 255.255.255.0
no shutdown
!
router IS-IS IGP
interface GigabitEthernet0/0/0/3
```

```
circuit-type level-2-only
point-to-point
hello-padding disable
address-family ipv4 unicast
 fast-reroute per-prefix
 fast-reroute per-prefix ti-lfa
 metric 10
```

Preparation for the tiebreaker scenarios is complete with the integration of router P9 into the topology, as described in the preceding configuration and depicted in *Figure 14.1*.

> **Exercise**
>
> Perform an exercise to validate IGP and SR-MPLS reachability from other routers to ensure that P9 is effectively integrated into the topology.

The next section covers the first tiebreaker scenario, namely **link protection preference**.

TI-LFA link protection preference

The TI-LFA link protection preference scenario is depicted in *Figure 14.2*.

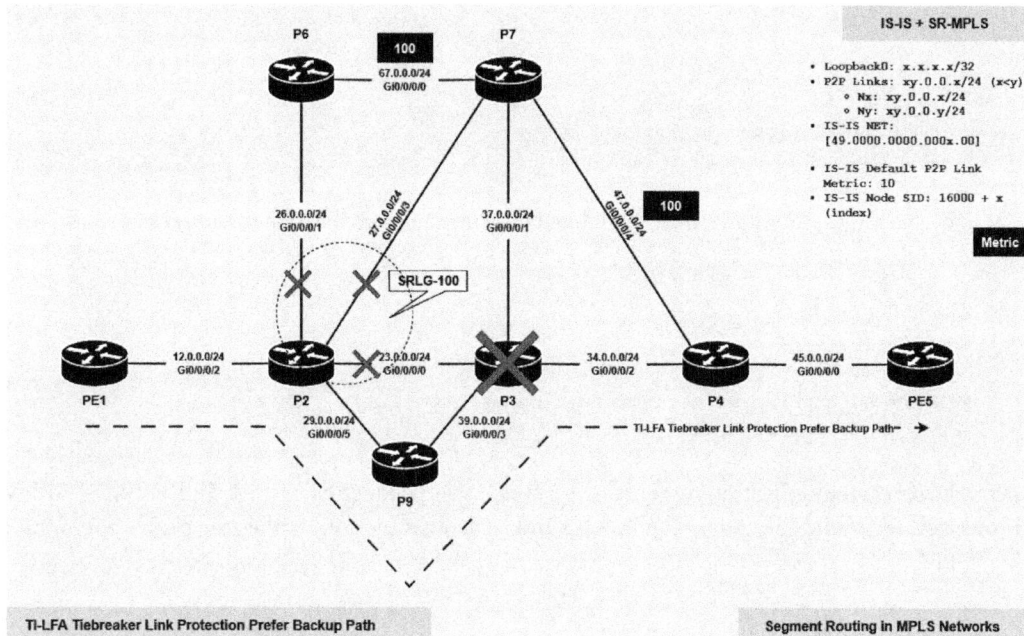

Figure 14.2 – TI-LFA tiebreaker link protection preference

As seen here, the topology is strategically designed to not offer all three TI-LFA path protections (link + node + SRLG) simultaneously.

Objective

The objective of this lab is to explore a scenario where the router requests TI-LFA to provide link (default) + node + SRLG protection. However, due to the absence of simultaneous protection and the nature of the topology, TI-LFA defaults to link protection preference. This objective is achieved through the tasks outlined as follows:

1.   Apply the required configuration to establish the topology, as illustrated in *Figure 14.2*.

2.   Verify that TI-LFA backup path protection prioritizes link protection.

Let's dive in.

Configuration

The following configuration adds the P2 - P6 link on the P2 router in the same SRLG group as the other two already configured links.

The SRLG-100 group is already configured on router P2, so the following configuration simply adds the interface toward router P6 to the same SRLG:

P2

```
srlg
interface GigabitEthernet0/0/0/1
  name SRLG-100
```

The node and SRLG protection remain configured on router P2 from the previous chapter, as shown here:

```
RP/0/RP0/CPU0:P2#show run router IS-IS

router IS-IS IGP
<..snipped..>
  fast-reroute per-prefix tiebreaker node-protecting index 200
  fast-reroute per-prefix tiebreaker srlg-disjoint index 100
```

With the preceding configuration, router P2 has essentially requested TI-LFA (link + node + SRLG) protection for traffic going over the P2 - P3 link. The next section verifies the protection offered by TI-LFA.

Verification

The following output indicates that the backup path is a classic **loop-free alternate** (**LFA**), as TI-LFA node protection and SRLG protection (node+SRLG) were not concurrently available, as depicted in *Figure 14.2*.

The path with node protection does not offer SRLG protection, and vice versa.

Backup path in RIB, FIB, and LFIB

The backup path offered is LFA or classic LFA, as shown in the following output:

```
RP/0/RP0/CPU0:P2#show IS-IS fast-reroute detail 5.5.5.5/32

L2 5.5.5.5/32 [30/115] Label: 16005, medium priority
   Installed Jan 01 08:01:33.244 for 00:00:14
     via 23.0.0.3, GigabitEthernet0/0/0/0, Label: 16005, P3, SRGB Base: 16000,
Weight: 0
       Backup path: LFA, via 29.0.0.9, GigabitEthernet0/0/0/5, Label: 16005,
P9, SRGB Base: 16000, Weight: 0, Metric: 40
       P: No, TM: 40, LC: No, NP: No, D: No, SRLG: Yes
     src PE5.00-00, 5.5.5.5, prefix-SID index 5, R:0 N:1 P:0 E:0 V:0 L:0,
Alg:0
RP/0/RP0/CPU0:P2#
```

The same is reflected in the routing table, as shown here:

```
RP/0/RP0/CPU0:P2#show route 5.5.5.5/32

Routing entry for 5.5.5.5/32
  Known via "IS-IS IGP", distance 115, metric 30, labeled SR, type level-2
  Installed Jan  1 08:01:33.244 for 00:00:28
  Routing Descriptor Blocks
    23.0.0.3, from 5.5.5.5, via GigabitEthernet0/0/0/0, Protected
      Route metric is 30
    29.0.0.9, from 5.5.5.5, via GigabitEthernet0/0/0/5, Backup (Local-LFA)
      Route metric is 40
  No advertising protos.
RP/0/RP0/CPU0:P2#
```

The FIB reflects the same information as the RIB, as shown in the following:

```
RP/0/RP0/CPU0:P2#show cef 5.5.5.5/32 brief
5.5.5.5/32, version 1118, labeled SR, internal 0x1000001 0x8310 (ptr
0xe7ac838) [1], 0x600 (0xda5df48), 0xa28 (0x2218f4f0)
Updated Jan  1 08:01:33.274
remote adjacency to GigabitEthernet0/0/0/0
Prefix Len 32, traffic index 0, precedence n/a, priority 1
   via 23.0.0.3/32, GigabitEthernet0/0/0/0, 10 dependencies, weight 0, class
```

```
0, protected [flags 0x400]
    path-idx 0 bkup-idx 1 NHID 0x0 [0xd674cc0 0x0]
    next hop 23.0.0.3/32
     local label 16005      labels imposed {16005}
    via 29.0.0.9/32, GigabitEthernet0/0/0/5, 21 dependencies, weight 0, class
0, backup (Local-LFA) [flags 0x300]
    path-idx 1 NHID 0x0 [0x22033550 0x0]
    next hop 29.0.0.9/32
    remote adjacency
     local label 16005      labels imposed {16005}
RP/0/RP0/CPU0:P2#
```

The LFIB also reflects the same information as the RIB, as shown here:

```
RP/0/RP0/CPU0:P2#show mpls forwarding prefix 5.5.5.5/32
Local  Outgoing    Prefix              Outgoing      Next Hop          Bytes
Label  Label       or ID               Interface                       Switched
------ ----------- ------------------- ------------- ---------------- ----------
--
16005  16005       SR Pfx (idx 5)      Gi0/0/0/0     23.0.0.3          0
       16005       SR Pfx (idx
5)     Gi0/0/0/5   29.0.0.9          0             (!)
RP/0/RP0/CPU0:P2#
```

The preceding output clearly indicates that neither node protection nor SRLG protection took precedence, and ultimately, link protection was preferred and offered by TI-LFA on router P2. The next section examines how this configuration came into effect.

Understanding backup path preference

The following examines each protection mechanism separately and describes the backup path route and the end-to-end cost through each path:

- **TI-LFA link protection**: This post-convergence path would be P2 - P9 – P3 – P4 – PE5, with a cost of 40.

    It would also be P2 – P7 – P3 – P4 – PE5, with a cost of 40 (but P2 – P7 has a local SRLG-disjoint, so it was not chosen).

- **TI-LFA node protection**: This post-convergence path would be P2 - P7 - P4 - PE5, with a cost of 120.

- **TI-LFA SRLG protection**: This post-convergence path would be P2 - P9 – P3 – P4 – PE5, with a cost of 40.

The TI-LFA link protection holds the lowest index (maximum preference) among the available paths when all protection methods are not simultaneously accessible.

> **Note**
>
> The TI-LFA link protection index cannot be modified due to platform restrictions.

Interestingly, another condition was also satisfied here: the classic LFA inequality condition explored earlier in the book, in *Chapters 5* and *6*.

As per RFC 5286, *Inequality 1: Loop-Free Criterion* is defined as a neighbor *N* that can provide an LFA if, and only if, `Distance_opt(N, D) < Distance_opt(N, S) + Distance_opt(S, D)`, where S is used to indicate the calculating router. `N_i` is a neighbor of S; N is used as an abbreviation when only one neighbor is being discussed. D is the destination under consideration.

In this case, `S = P2`, `D = PE5`, and `N = P9` (as it has the shortest path among all the neighbors).

Hence, the inequality condition for P9 is as follows:

```
S = P2, D = PE5 and N = P9
Distance_opt(P9, PE5) < Distance_opt(P9, P2) + Distance_opt(P2, PE5)
30 (P3 - P4 - PE5) < 10 + 30 (P3 - P4 - PE5)
```

The condition is satisfied.

> **Note**
>
> The inequality condition is satisfied for the neighbor P7 as well, but the PLR P2 selected P9 as the next hop because P7 also violated the local SRLG-disjoint, and was therefore not chosen.

The TI-LFA link protection backup path is identical to the classic LFA backup path. Once an alternative path is discovered through classic LFA, it becomes the preferred choice over the TI-LFA link protection method, as indicated in the preceding output and seen in *Chapters 5* and *6*.

Therefore, TI-LFA link protection was offered as the preferred fast reroute mechanism. The next section covers the SRLG Protection Preference scenario.

TI-LFA SRLG protection preference

The TI-LFA SRLG protection preference scenario is depicted in *Figure 14.3*.

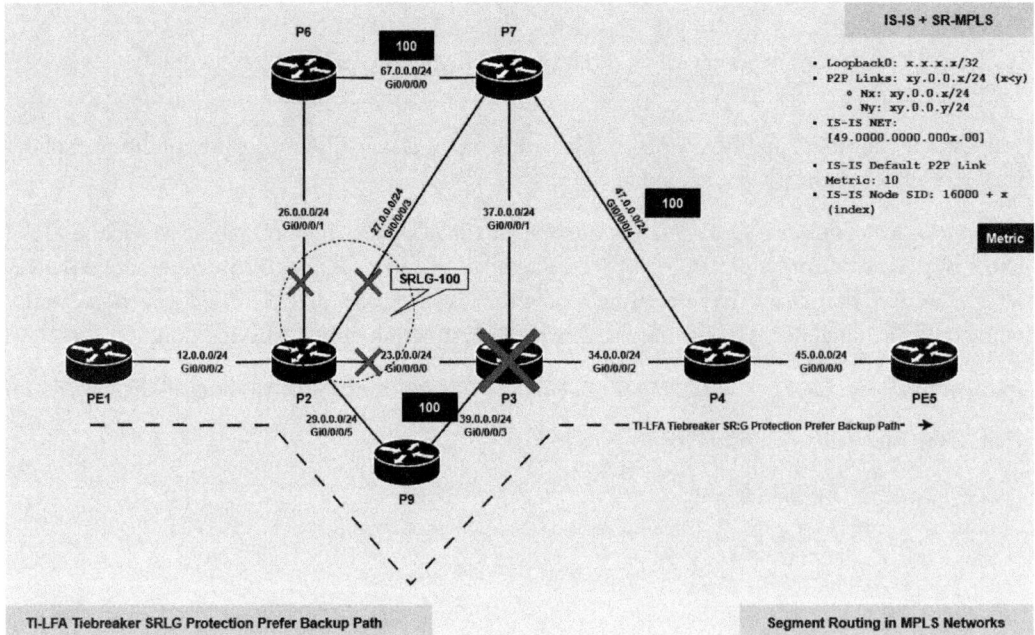

Figure 14.3 – TI-LFA tiebreaker SRLG protection preference

As seen in the preceding figure, the topology is strategically designed to not offer all three TI-LFA path protections (link + node + SRLG) simultaneously.

Objective

The objective of this lab is to explore a scenario where the router requests TI-LFA to provide link (default) + node + SRLG protection. However, due to the absence of simultaneous protection and the nature of the topology, TI-LFA prefers SRLG protection. This objective is achieved through the tasks outlined here:

1. Apply the required configuration to establish the topology as illustrated in *Figure 14.3*.

2. Verify that TI-LFA backup path protection prioritizes SRLG protection.

Let's dive in.

Configuration

Increase the metric on P9 toward P3 to ensure that neither the classic LFA nor the TI-LFA link protection backup path is accessible from P2 to PE5:

P9

```
router IS-IS IGP
interface GigabitEthernet0/0/0/3
  address-family ipv4 unicast
   metric 100
```

Increasing the metric on router P9 toward router P3 dissatisfied the inequality condition as seen in the previous lab, *TI-LFA link protection preference*.

With this change, router P9 itself uses the P2-P3 link to reach router PE5, incurring additional metrics on the end-to-end path cost, eventually dissatisfying the inequality condition.

The protection offered by TI-LFA in this scenario will be covered in the next section.

Verification

The backup path is examined in the RIB, FIB, and LFIB here to verify the offered protection mechanism.

Backup path in RIB, FIB, and LFIB

The following output confirms that SRLG protection is offered:

```
RP/0/RP0/CPU0:P2#show IS-IS fast-reroute detail 5.5.5.5/32


L2 5.5.5.5/32 [30/115] Label: 16005, medium priority
    Installed Jan 01 08:04:37.559 for 00:00:06
      via 23.0.0.3, GigabitEthernet0/0/0/0, Label: 16005, P3, SRGB Base: 16000,
Weight: 0
        Backup path: TI-LFA (srlg), via 29.0.0.9, GigabitEthernet0/0/0/5 P9,
SRGB Base: 16000, Weight: 0, Metric: 130
          P node: P9.00 [9.9.9.9], Label: ImpNull
          Q node: P3.00 [3.3.3.3], Label: 24003
          Prefix label: 16005
          Backup-src: PE5.00
         P: No, TM: 130, LC: No, NP: No, D: No, SRLG: Yes
      src PE5.00-00, 5.5.5.5, prefix-SID index 5, R:0 N:1 P:0 E:0 V:0 L:0,
Alg:0
RP/0/RP0/CPU0:P2#
```

The same is reflected in the routing table, as shown here:

```
RP/0/RP0/CPU0:P2#show route 5.5.5.5/32

Routing entry for 5.5.5.5/32
  Known via "IS-IS IGP", distance 115, metric 30, labeled SR, type level-2
  Installed Jan  1 08:04:37.560 for 00:00:22
  Routing Descriptor Blocks
    23.0.0.3, from 5.5.5.5, via GigabitEthernet0/0/0/0, Protected
      Route metric is 30
    29.0.0.9, from 5.5.5.5, via GigabitEthernet0/0/0/5, Backup (TI-LFA)
      Repair Node(s): 9.9.9.9, 3.3.3.3
      Route metric is 130
  No advertising protos.
RP/0/RP0/CPU0:P2#
```

The FIB downloads the same information from the RIB.

The SRLG protection is offered via a repair tunnel through routers P9 and P3, with the label stack shown in the following output:

```
RP/0/RP0/CPU0:P2#show cef 5.5.5.5/32 brief
5.5.5.5/32, version 1147, labeled SR, internal 0x1000001 0x8310 (ptr
0xe7ac838) [1], 0x600 (0xda5ef50), 0xa28 (0x2218f1d8)
Updated Jan  1 08:04:37.576
remote adjacency to GigabitEthernet0/0/0/0
Prefix Len 32, traffic index 0, precedence n/a, priority 1
  via 23.0.0.3/32, GigabitEthernet0/0/0/0, 14 dependencies, weight 0, class
0, protected [flags 0x400]
    path-idx 0 bkup-idx 1 NHID 0x0 [0xd674cc0 0xd674cc0]
    next hop 23.0.0.3/32
     local label 16005      labels imposed {16005}
  via 29.0.0.9/32, GigabitEthernet0/0/0/5, 22 dependencies, weight 0, class
0, backup (TI-LFA) [flags 0xb00]
    path-idx 1 NHID 0x0 [0x22033550 0x0]
    next hop 29.0.0.9/32, Repair Node(s): 9.9.9.9, 3.3.3.3
    remote adjacency
     local label 16005      labels imposed {ImplNull 24003 16005}
RP/0/RP0/CPU0:P2#
```

The LFIB also downloads the same information from the RIB, as shown here:

```
RP/0/RP0/CPU0:P2#show mpls forwarding prefix 5.5.5.5/32 detail
Local  Outgoing    Prefix             Outgoing      Next Hop        Bytes
Label  Label       or ID              Interface                     Switched
------ ----------- ------------------ ------------ --------------- ----------
--
16005  16005       SR Pfx (idx 5)     Gi0/0/0/0    23.0.0.3        0
```

```
          Updated: Jan  1 08:04:37.576
          Path Flags: 0x400 [  BKUP-IDX:1 (0xd674cc0) ]
          Version: 1147, Priority: 1
          Label Stack (Top -> Bottom): { 16005 }
          NHID: 0x0, Encap-ID: N/A, Path idx: 0, Backup path idx: 1, Weight: 0
          MAC/Encaps: 4/8, MTU: 1500
          Outgoing Interface: GigabitEthernet0/0/0/0 (ifhandle 0x01000020)
          Packets Switched: 0


          Pop         SR Pfx (idx
  5)      Gi0/0/0/5   29.0.0.9         0              (!)
          Updated: Jan  1 08:04:37.576
          Path Flags: 0xb00 [  IDX:1 BKUP, NoFwd ]
          Version: 1147, Priority: 1
          Label Stack (Top -> Bottom): { Imp-Null 24003 16005 }
          NHID: 0x0, Encap-ID: N/A, Path idx: 1, Backup path idx: 0, Weight: 0
          MAC/Encaps: 4/12, MTU: 1500
          Outgoing Interface: GigabitEthernet0/0/0/5 (ifhandle 0x01000038)
          Packets Switched: 0
          (!): FRR pure backup

    Traffic-Matrix Packets/Bytes Switched: 0/0
  RP/0/RP0/CPU0:P2#
```

The preceding output indicates that SRLG protection was preferred and offered by TI-LFA on router P2. The next section examines how this configuration came into effect.

Understanding backup path preference

The following examines each protection mechanism separately and describes the backup path route and the end-to-end cost through each path:

- **TI-LFA node protection**: This post-convergence path would be P2 - P7 - P4 - PE5, with a cost of 120
- **TI-LFA SRLG protection**: This post-convergence path would be P2 - P9 – P3 – P4 – PE5, with a cost of 130

The TI-LFA node protection backup path has a lower cost compared to the SRLG protection backup path. However, the SRLG protection tiebreaker index is 100, whereas the node protection index is 200.

The lowest index holds maximum preference, and since link protection is not available, the SRLG protection backup path is chosen, as depicted in the preceding output.

Therefore, TI-LFA SRLG protection was offered as the preferred fast reroute mechanism. The next section covers the node protection preference scenario.

TI-LFA node protection preference

The TI-LFA node protection preference scenario is depicted in *Figure 14.4*.

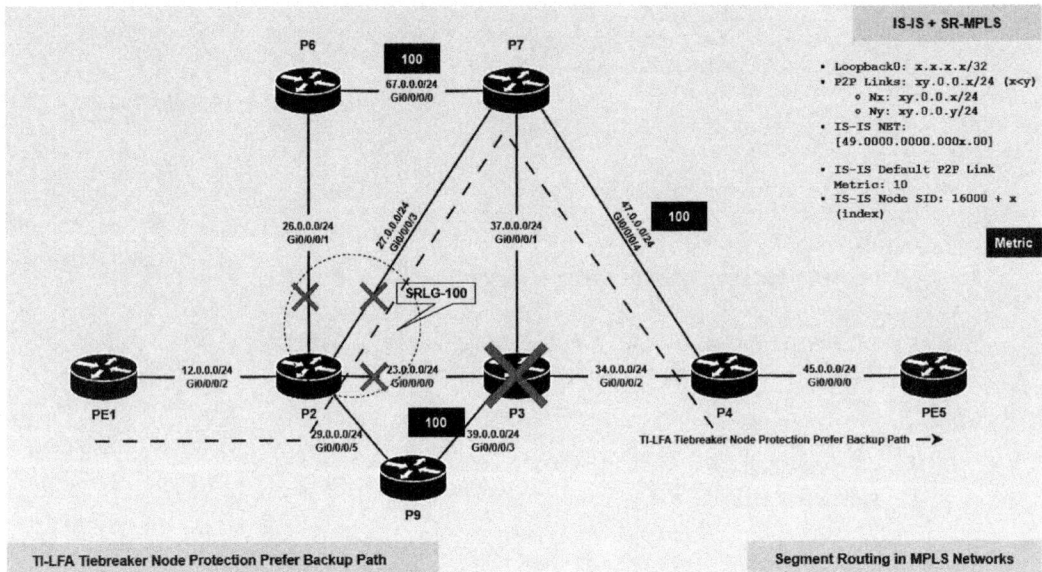

Figure 14.4 – TI-LFA tiebreaker node protection preference

As seen here, the topology is strategically designed to not offer all three TI-LFA path protections (link + node + SRLG) simultaneously.

Objective

The objective of this lab is to explore a scenario where the router requests TI-LFA to provide link (default) + node + SRLG protection. However, due to the absence of simultaneous protection and the nature of the topology, TI-LFA prefers node protection. This objective is achieved through the tasks outlined here:

1. Prefer node protection over SRLG protection on P2 by increasing the SRLG index.
2. Verify that TI-LFA backup path protection prioritizes node protection.

Let's dive in.

Configuration

The lower the tiebreaker index, the higher the protection preference.

Raise the SRLG protection tiebreaker index to a value exceeding 200 to make the TI-LFA backup path prioritize node protection when only one of them can be used, but not both:

P2

```
router IS-IS IGP
address-family ipv4 unicast
  fast-reroute per-prefix tiebreaker srlg-disjoint index 255
```

The node protection index is 200, as configured in the previous chapter. The SRLG protection index is increased to 255, making node protection the preferred choice when both cannot be offered simultaneously.

The protection offered by TI-LFA in this scenario will be covered in the next section.

Verification

The backup path is examined in the RIB, FIB, and LFIB to verify the offered protection mechanism.

Backup path in RIB, FIB, and LFIB

The following output confirms that node protection is offered:

```
RP/0/RP0/CPU0:P2#show IS-IS fast-reroute detail 5.5.5.5/32

L2 5.5.5.5/32 [30/115] Label: 16005, medium priority
    Installed Jan 01 08:07:36.907 for 00:00:12
      via 23.0.0.3, GigabitEthernet0/0/0/0, Label: 16005, P3, SRGB Base: 16000,
Weight: 0
        Backup path: TI-LFA (node), via 27.0.0.7, GigabitEthernet0/0/0/3 P7,
SRGB Base: 16000, Weight: 0, Metric: 120
          P node: P7.00 [7.7.7.7], Label: ImpNull
          Q node: P4.00 [4.4.4.4], Label: 24023
          Prefix label: 16005
          Backup-src: PE5.00
        P: No, TM: 120, LC: No, NP: Yes, D: No, SRLG: No
      src PE5.00-00, 5.5.5.5, prefix-SID index 5, R:0 N:1 P:0 E:0 V:0 L:0,
Alg:0
RP/0/RP0/CPU0:P2#
```

The same is reflected in the routing table, as shown here:

```
RP/0/RP0/CPU0:P2#show route 5.5.5.5/32 detail

Routing entry for 5.5.5.5/32
  Known via "IS-IS IGP", distance 115, metric 30, labeled SR, type level-2
  Installed Jan  1 08:07:36.907 for 00:00:31
```

```
    Routing Descriptor Blocks
      23.0.0.3, from 5.5.5.5, via GigabitEthernet0/0/0/0, Protected
        Route metric is 30
        Label: 0x3e85 (16005)
        Tunnel ID: None
        Binding Label: None
        Extended communities count: 0
        Path id:1       Path ref count:0
        NHID:0x7(Ref:10)
        Backup path id:65
      27.0.0.7, from 5.5.5.5, via GigabitEthernet0/0/0/3, Backup (TI-LFA)
        Repair Node(s): 7.7.7.7, 4.4.4.4
        Route metric is 120
        Labels: 0x3 0x5dd7 0x3e85 (3 24023 16005)
        Tunnel ID: None
        Binding Label: None
        Extended communities count: 0
        Path id:65              Path ref count:1
        NHID:0x9(Ref:11)
    Route version is 0x59 (89)
    Local Label: 0x3e85 (16005)
    IP Precedence: Not Set
    QoS Group ID: Not Set
    Flow-tag: Not Set
    Fwd-class: Not Set
    Route Priority: RIB_PRIORITY_NON_RECURSIVE_MEDIUM (7) SVD Type RIB_SVD_TYPE_
LOCAL
    Download Priority 1, Download Version 1164
    No advertising protos.
RP/0/RP0/CPU0:P2#
```

The FIB downloads the same information from the RIB.

The node protection is offered via a repair tunnel through routers P7 and P4, with the label stack shown in the following output:

```
RP/0/RP0/CPU0:P2#show cef 5.5.5.5/32 brief
5.5.5.5/32, version 1164, labeled SR, internal 0x1000001 0x8310 (ptr
0xe7ac838) [1], 0x600 (0xda5ef50), 0xa28 (0x2218f910)
Updated Jan  1 08:07:36.922
remote adjacency to GigabitEthernet0/0/0/0
Prefix Len 32, traffic index 0, precedence n/a, priority 1
   via 23.0.0.3/32, GigabitEthernet0/0/0/0, 8 dependencies, weight 0, class 0,
protected [flags 0x400]
     path-idx 0 bkup-idx 1 NHID 0x0 [0xd673eb0 0xd673eb0]
     next hop 23.0.0.3/32
```

```
         local label 16005        labels imposed {16005}
      via 27.0.0.7/32, GigabitEthernet0/0/0/3, 8 dependencies, weight 0, class 0,
   backup (TI-LFA) [flags 0xb00]
         path-idx 1 NHID 0x0 [0x220335f0 0x0]
         next hop 27.0.0.7/32, Repair Node(s): 7.7.7.7, 4.4.4.4
         remote adjacency
         local label 16005        labels imposed {ImplNull 24023 16005}
   RP/0/RP0/CPU0:P2#
```

The LFIB also downloads the same information from the RIB, as shown here:

```
RP/0/RP0/CPU0:P2#show mpls forwarding prefix 5.5.5.5/32 detail
Local  Outgoing    Prefix              Outgoing      Next Hop          Bytes
Label  Label       or ID               Interface                       Switched
------ ----------- ------------------- ------------- ----------------- ----------
--
16005  16005       SR Pfx (idx 5)      Gi0/0/0/0     23.0.0.3          0
       Updated: Jan  1 08:07:36.923
       Path Flags: 0x400 [  BKUP-IDX:1 (0xd673eb0) ]
       Version: 1164, Priority: 1
       Label Stack (Top -> Bottom): { 16005 }
       NHID: 0x0, Encap-ID: N/A, Path idx: 0, Backup path idx: 1, Weight: 0
       MAC/Encaps: 4/8, MTU: 1500
       Outgoing Interface: GigabitEthernet0/0/0/0 (ifhandle 0x01000020)
       Packets Switched: 0

       Pop         SR Pfx (idx
5)     Gi0/0/0/3    27.0.0.7           0             (!)
       Updated: Jan  1 08:07:36.923
       Path Flags: 0xb00 [  IDX:1 BKUP, NoFwd ]
       Version: 1164, Priority: 1
       Label Stack (Top -> Bottom): { Imp-Null 24023 16005 }
       NHID: 0x0, Encap-ID: N/A, Path idx: 1, Backup path idx: 0, Weight: 0
       MAC/Encaps: 4/12, MTU: 1500
       Outgoing Interface: GigabitEthernet0/0/0/3 (ifhandle 0x01000048)
       Packets Switched: 0
       (!): FRR pure backup

   Traffic-Matrix Packets/Bytes Switched: 0/0
   RP/0/RP0/CPU0:P2#
```

The preceding output indicates that node protection was preferred and offered by TI-LFA on router P2. The next section examines how this configuration came into effect.

Understanding backup path preference

The following examines each protection mechanism separately and describes the backup path route and the end-to-end cost through each path:

- **TI-LFA node protection**: This post-convergence path would be P2 - P7 - P4 - PE5, with a cost of 120

- **TI-LFA SRLG protection**: This post-convergence path would be P2 - P9 - P3 - P4 - PE5, with a cost of 130

The SRLG protection tiebreaker index is 255, whereas the node protection index is 200.

The lowest index holds maximum preference, and since link protection is not available, the node protection backup path is chosen, as depicted in the preceding output.

Therefore, TI-LFA node protection was offered as the preferred fast reroute mechanism.

This concludes the final chapter of this book.

Summary

This chapter builds on the comprehensive TI-LFA node + SRLG protection scenario presented in the previous chapter, with the objective of exploring scenarios where TI-LFA cannot offer all the requested protection mechanisms simultaneously. This situation requires the PLR to decide which protection mechanism takes precedence.

The Cisco IOS-XR platform allows for the configuration of node and SRLG protection using tiebreaker index values. Although these index values were introduced in previous chapters, their significance is emphasized in this chapter. The index values help TI-LFA on a router determine which protection mechanism to prefer when not all requested protections can be offered simultaneously. A lower index indicates a higher preference for that protection.

To explore the tiebreaker scenarios, an additional router, P9, was integrated into the existing topology. The configuration required to integrate P9 adheres to the same principles used for other routers, including the IPv4 address schema, IS-IS NET-ID, and SR-MPLS Prefix-SID index introduced in *Chapters 2* and *3*. The strategic placement of router P9 was designed to explore all three intended protection scenarios—link protection preference, SRLG protection preference, and node protection preference—in sequence. This setup ensured that TI-LFA could not offer all three protections simultaneously, and each scenario's preferred path was validated using the RIB, FIB, and LFIB outputs. In each scenario, backup paths were individually assessed to determine why all protections could not be offered simultaneously and how each protection method was preferred over the others.

The TI-LFA scenarios, along with SR-MPLS and SR-LDP interworking, were included to teach the concept of source or segment routing through multiple scenarios and to demonstrate fast reroute protection mechanisms in SR-MPLS networks. Throughout the labs, emphasis was placed on routing determined at source or PLR for fast reroute, without maintaining any state in the network, as the packet itself carried the state through the configuration of the label stack, whether for the explicit primary path in the microloop avoidance lab or the backup path in the TI-LFA labs.

With the conclusion of this chapter, the book's objective is also fulfilled. The goal was to make it easier for readers to learn the concepts of MPLS using IS-IS as the IGP and then to venture into SR-MPLS networks with TI-LFA fast reroute protection. While there is a lot more to segment routing than what is covered in this book, the purpose was to provide readers with a solid introduction to the world of segment routing.

References

Below are the key references cited throughout this chapter. These documents provide additional insights and technical details on MPLS, Fast Reroute, and Segment Routing, offering further reading for those interested in exploring these topics in depth.

- [RFC5286] Atlas, A., Ed. and A. Zinin, Ed., "Basic Specification for IP Fast Reroute: Loop-Free Alternates", RFC 5286, DOI 10.17487/RFC5286, September 2008, <https://www.rfc-editor.org/info/rfc5286>.

- [RFC7490] Bryant, S., Filsfils, C., Previdi, S., Shand, M., and N. So, "Remote Loop-Free Alternate (LFA) Fast Reroute (FRR)", RFC 7490, DOI 10.17487/RFC7490, April 2015, <https://www.rfc-editor.org/info/rfc7490>.

- [RFC7916] Litkowski, S., Ed., Decraene, B., Filsfils, C., Raza, K., Horneffer, M., and P. Sarkar, "Operational Management of Loop-Free Alternates", RFC 7916, DOI 10.17487/RFC7916, July 2016, <https://www.rfc-editor.org/info/rfc7916>.

- [RFC8029] Kompella, K., Swallow, G., Pignataro, C., Ed., Kumar, N., Aldrin, S., and M. Chen, "Detecting Multiprotocol Label Switched (MPLS) Data-Plane Failures", RFC 8029, DOI 10.17487/RFC8029, March 2017, <https://www.rfc-editor.org/info/rfc8029>.

- [I-D.ietf-rtgwg-segment-routing-ti-lfa] Bashandy, A., Litkowski, S., Filsfils, C., Francois, P., Decraene, B., and D. Voyer, "Topology Independent Fast Reroute using Segment Routing", Work in Progress, Internet-Draft, draft-ietf-rtgwg-segment-routing-ti-lfa-13, 16 January 2024, <https://datatracker.ietf.org/doc/html/draft-ietf-rtgwg-segment-routing-ti-lfa-13>.

- [RFC5715] Shand, M. and S. Bryant, "A Framework for Loop-Free Convergence", RFC 5715, DOI 10.17487/RFC5715, January 2010, <https://www.rfc-editor.org/info/rfc5715>.

- [RFC6571] Filsfils, C., Ed., Francois, P., Ed., Shand, M., Decraene, B., Uttaro, J., Leymann, N., and M. Horneffer, "Loop-Free Alternate (LFA) Applicability in Service Provider (SP) Networks", RFC 6571, DOI 10.17487/RFC6571, June 2012, <https://www.rfc-editor.org/info/rfc6571>.

Index

‹packt›

packtpub.com

Subscribe to our online digital library for full access to over 7,000 books and videos, as well as industry leading tools to help you plan your personal development and advance your career. For more information, please visit our website.

Why subscribe?

- Spend less time learning and more time coding with practical eBooks and Videos from over 4,000 industry professionals

- Improve your learning with Skill Plans built especially for you

- Get a free eBook or video every month

- Fully searchable for easy access to vital information

- Copy and paste, print, and bookmark content

Did you know that Packt offers eBook versions of every book published, with PDF and ePub files available? You can upgrade to the eBook version at packtpub.com and as a print book customer, you are entitled to a discount on the eBook copy. Get in touch with us at customercare@packtpub.com for more details.

At www.packtpub.com, you can also read a collection of free technical articles, sign up for a range of free newsletters, and receive exclusive discounts and offers on Packt books and eBooks.

Other Books You May Enjoy

If you enjoyed this book, you may be interested in these other books by Packt:

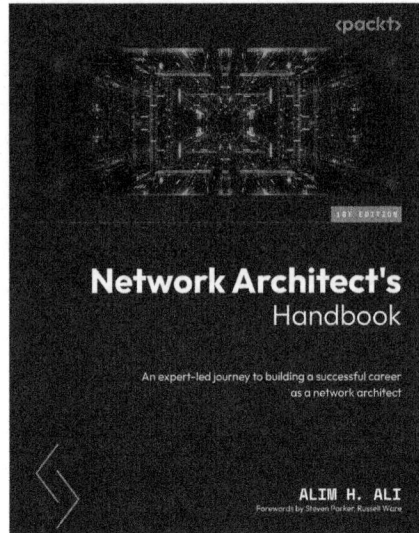

Network Architect's Handbook

Alim H. Ali

ISBN: 978-1-83763-783-6

- Examine the role of a network architect
- Understand the key design makers in an organization
- Choose the best strategies to meet stakeholder needs
- Be well-versed with networking concepts
- Prepare for a network architect position interview
- Distinguish the different IT architects in an organization
- Identify relevant certification for network architects
- Understand the various de facto network/fabric architect models used today

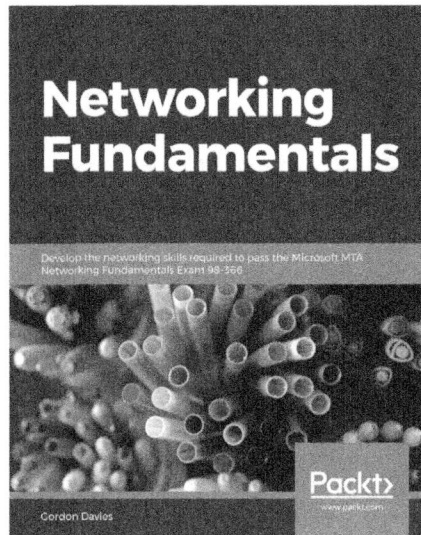

Networking Fundamentals

Gordon Davies

ISBN: 978-1-83864-350-8

- Things you will learn:
- Become well versed in networking topologies and concepts
- Understand network infrastructures such as intranets, extranets, and more
- Explore network switches, routers, and other network hardware devices
- Get to grips with different network protocols and models such as OSI and TCP/IP
- Work with a variety of network services such as DHCP, NAT, firewalls, and remote access
- Apply networking concepts in different real-world scenarios

Packt is searching for authors like you

If you're interested in becoming an author for Packt, please visit authors.packtpub.com and apply today. We have worked with thousands of developers and tech professionals, just like you, to help them share their insight with the global tech community. You can make a general application, apply for a specific hot topic that we are recruiting an author for, or submit your own idea.

Share Your Thoughts

Now you've finished *Segment Routing in MPLS Networks*, we'd love to hear your thoughts! Scan the QR code below to go straight to the Amazon review page for this book and share your feedback or leave a review on the site that you purchased it from.

https://packt.link/r/1-836-20321-7

Your review is important to us and the tech community and will help us make sure we're delivering excellent quality content.

Download a free PDF copy of this book

Thanks for purchasing this book!

Do you like to read on the go but are unable to carry your print books everywhere?

Is your eBook purchase not compatible with the device of your choice?

Don't worry, now with every Packt book you get a DRM-free PDF version of that book at no cost.

Read anywhere, any place, on any device. Search, copy, and paste code from your favorite technical books directly into your application.

The perks don't stop there, you can get exclusive access to discounts, newsletters, and great free content in your inbox daily

Follow these simple steps to get the benefits:

1. Scan the QR code or visit the link below

https://packt.link/free-ebook/978-1-83620-321-6

2. Submit your proof of purchase
3. That's it! We'll send your free PDF and other benefits to your email directly

Printed in Great Britain
by Amazon